AF394362

PROPHECY

PROPHECY

Prediction, Power, and the Fight for the Future, from Ancient Oracles to AI

CARISSA VÉLIZ

Swift

SWIFT PRESS

First published in Great Britain by Swift Press 2026
First published in the United States of America
by Doubleday, a division of Penguin Random House LLC

1 3 5 7 9 8 6 4 2

Page 28 illustration: Wikimedia Commons
Page 46 photograph: Zuri Swimmer / Alamy
Page 170 photograph: David Ramos / Getty Images
Page 251 photograph: Anadolu / Getty Images

Printed and bound in Great Britain by CPI Group (UK) Ltd, Croydon CRO 4YY

A CIP catalogue record for this book is available from the British Library

We make every effort to make sure our products are safe for the purpose for which
they are intended. Our authorised representative in the EU for product safety is
Easy Access System Europe, Mustamäe tee 50, 10621 Tallinn, Estonia
gpsr.requests@easproject.com

ISBN: 9781800757240
eISBN: 9781800757257

A mi madre, siempre a mi madre

And to Maia, Emilia, and Sara,

who make me feel that the future is bright

Divination is a gift of God.
Therefore its abuse should be
treated as a punishable imposture.

—Michel de Montaigne, *Essays*

CONTENTS

PART THREE: RETHINKING PREDICTION

PRELUDE

What if I told you that a prophecy led you to this book? It's meant for you. It's in your cards to read it as much as it was in mine to write it.

Books have ways of choosing their readers. Maybe an algorithmic prediction identified you as someone who will enjoy this book and showed you an advertisement for it. Or perhaps your local bookseller or librarian knows you well enough to foresee that these pages are for you. Or you might be one of the lucky ones who has a friend with a talent for suggesting just the right title.

Regardless of how this book found you, a series of forecasts lies behind its journey. The predictions of the author, agent, editors, publicists, marketers, and algorithms have weighed in on its life.

Predictions are what connected you and me. Our minds meet through prophecy.

Your life story is not all that different from that of this book. Countless predictions have directed your path too, opening and closing doors for you, pushing you to some places while blocking the way to others, leading you to where you are now.

As you read these sentences, artificial intelligence is making forecasts about you. Algorithms are deciding whether you are eligible for a loan, a job, an apartment, or insurance. They determine what you see online, who reads your social media posts, and who connects with you on dating apps. They may even decide whether you get arrested or go to jail. Your very life hangs in the balance of prophecies. If you're unlucky, a prediction will be what kills you. Forecasts can determine your place in a waiting list for an organ transplant, or whether you get medical care in an emergency. Policymaking hinges on predictions. War and peace, and whether someone lives or dies, are decided based on forecasts about the strength of an adversary, the impact of a mission, or the identity of a person.

And yet no one has asked your permission to make those guesses. No governmental agency is supervising them. No one is informing you of the prophecies that shape your fate.

Prophecies are the grounds on which fights over the future take place. Our expectations bend the social world toward our predictions. When someone forecasts that the world will be a certain way, they are commanding that others obey their wishes and bring that world about. Even though we have been using predictions for thousands of years to make some of the most important decisions of our lives, we have dedicated remarkably little thought to the deeper questions about prophecy: What exactly are predictions, what are their effects, who has the authority to make them, and when is it appropriate to use them?

Prediction is at the heart of what it is to be human. Part of what makes us strong despite our weak constitutions—along with opposable thumbs and the power of language—is our ability to foretell events. It gives us a competitive edge and has helped us survive for hundreds of thousands of years.

When we can foresee the behavior of other animals, it's easier to hunt them. When we can forecast what the next season will be like, we can prepare for it. When we can estimate the movement of the stars, we can use the sky as a watch, a calendar, and a map. It is unsurprising that we put so much hope on prediction. It's gotten us this far. If you are running a company, everything from choosing your line of business to finding the right location for it and deciding how to market your product depends on forecasting.

But prediction is a double-edged sword. It can be the harbinger of our success, or of our downfall. When Croesus, the king of Lydia, asked the Oracle of Delphi if he should attack Persia, the Oracle responded, "If you cross the river, a great empire will be destroyed."

He interpreted this forecast as a good omen and invaded Persia. A great empire was indeed destroyed—his own.

We are vulnerable to prediction because we are wishful, anxious creatures who crave for certainty. One of our greatest assets is our imagination, which allows us to dream up better worlds and pursue them. But that same imagination strikes terror into us, keeping us up at night, wondering about what awaits us. Will you have a job in the future? Will your family stay together, safe, and healthy? Will your company thrive? Will your country remain free? Will war break out? Will climate change be the end of us?

We know that terrible things happen to human beings every second of every day. You've seen them happen to your friends, or your neighbors, or your colleagues. Maybe something horrific has happened to you, and your body remembers. Some of us get painful and sudden diseases, schools and hospitals get bombed, vulnerable people get raped, writers get knifed, dissidents and minorities get persecuted, cars and airplanes crash spectacularly. Sometimes I think of people as walking ketchup packets, the only thing between us and disintegration being a thin layer of permeable skin. Just slightly too much pressure, and we go *splat*.

The world is a ruthless, dangerous, and beautiful place to inhabit. In the sound of your every step a question echoes: Is today the day when it all falls apart? I begin my philosophy classes by reminding my students that, statistically, their lives are likely to crumble to rubble at least twice.

We all care about the future because that's where we'll be spending the rest of our lives.[1] But our anxiety about tomorrow makes us vulnerable to charlatans, dodgy technology, and self-deception. We seek out predictions because we desperately want to be reassured that it's going to be alright. But putting ourselves in the hands of prophets perversely makes us more unsafe and less free. The more we understand how prediction works, and the more we come to terms with uncertainty, the less likely we are to fall for the tricks of prophets.

—

This is a book that, I hope, will be hard to classify, because part of its purpose is to point out the limits of classification. Prediction is a topic so extensive that any book aspiring to do it justice must necessarily be broad in scope.

This is a technology book, because artificial intelligence is the new Oracle of Delphi and tech executives the new prophets. It's a business book, and a personal development one, because it explores prophecy as an industry, with a focus on how to avoid falling for false prophets in a "data-driven" world. It explores the history of prediction—from ancient oracles to the development of measurement and probability—to argue that prophecies are more often than not power plays in disguise. It's a political treatise on how prediction serves social control.

It's a book about climate change and other major challenges and how to tackle them. It's a defense of great literature and a look at the publishing industry. It's a criticism of the worst of academia, and a defense of the best of it. It has glimpses of a memoir, because it gives a sense of what it's like to be a professor at Oxford (it's wild; ceilings collapse, and predation and greatness abound). I appreciate it when authors reveal where they're coming from, and I didn't want to cover my tracks by fabricating a pretended view from nowhere.

It's a philosophy book, an exploration of what we say when we utter a statement about the future, and a reflection on the value of virtue, truth, and beauty. It's a book about creativity and humor, with some great jokes by comedic geniuses and some terrible ones by me (because if they were all good, you wouldn't appreciate them).

But most of all, it's a book about how to live well. It's about how the good life is not a script to discover or follow, but a blank page to write on.

OVERTURE

What's in a Prediction?

Forecasts could hardly be more pervasive or consequential in our lives, and yet they are grossly misunderstood and consistently misinterpreted—sometimes innocently, often negligently, and other times maliciously. What exactly is a prediction?

THE WISE VIEW OF PREDICTION

To make a prediction is to say or estimate that something will happen in the future. I use "prediction," "forecast," and "prophecy" as rough synonyms.

The most common misconception about forecasting is what I call *the naive view of prediction*. The naive view portrays predictions as quests for truth or a kind of knowledge. When tech companies sell predictive algorithms to governments and businesses, they sell the story that predictions will help them understand their citizens or clients, that they will allow them to anticipate problems. What the naive view misses are five crucial characteristics of predictions.

First, *predictions are guesses*.

Predictions are not facts. Facts belong to the present and the past. An assertion about the future can be many things—an estimate, a desire, a warning—but never a fact. What makes the future the future is that it's not yet here, it hasn't yet happened. What hasn't come to pass doesn't exist, and there are no facts about what doesn't exist.

You might think that picture must be wrong because it seems as if you can lie about the future, and if you can lie, then you must be able to tell the truth about it as well. But that's an illusion. You can lie about your present *beliefs* about the future, but not about the future itself; you can be wrong about the future, but if you were being sincere, it wasn't a lie.

That doesn't mean that all predictions are unrelated to facts. Some forecasts are justified. When 60 percent of the times that a weather app says that there is a 60 percent chance of rain, it rains, then we have good reason to trust it. But most predictions out there are not like weather forecasts.

Our justice system is supposed to be based on what people deserve. You don't get punished for anything you haven't done—in theory. Our educational system is supposed to be based on merit. You don't get grades for exams you haven't sat—in theory. Our job market is supposed to be fair. You don't get penalized for blunders you haven't made—in theory. What people have or haven't done are facts. When we use prediction to make decisions about human beings, we treat them according to guesses, not facts. We punish not for what someone did but for what *we* think someone will do. And we don't even give people the chance to dispute these predictions, because they are unverifiable and unfalsifiable: Since they are about the future, predictions cannot be challenged for being false, creating yet another injustice and allowing prophets to evade accountability.[1]

Second, *predictions are wishful.*

What we desire influences our predictions. Probability scholars have long recognized two aspects of prediction: a subjective, or epistemic aspect (from the Greek *epistēmē,* meaning knowledge or understanding), related to what we know; and an objective one, related to how the world is, irrespective of our knowledge. But they have largely ignored how our own desires shape our forecasts.

Any decision concerning *risk* also involves objective facts and epistemic views. What we think we know affects the risks we are willing to take. But it's not only about knowledge. It might be that you and I have the same knowledge about two horses, and it's reasonable to think that it's going to be a tight race, so it would be risky to bet on either one of them. But you might have a larger appetite for risk than

I do—perhaps because you are wealthier or enjoy adrenaline more—making you more willing to place a bet.

Our differences in judging and tolerating risk are essential for our society. There would be no stock market without variety in risk assessments. Without the venturesome, there would be less innovation and entrepreneurship. Without the cautious, the world would be more unstable and dangerous.

What is missing from our thinking about prediction is a third aspect of both probability and risk assessment related to our desires, and the main protagonist of this book: *the wishful aspect of prediction*. When you are predicting which horse will win a race, you are likely not a neutral observer. You *want* the horse you bet on to win.

Every person who starts a company thinks it will do well (money launderers and self-saboteurs aside). Your prediction is not dispassionate. You *crave* that it will go well, and you may want it so badly that your desire affects not only your risk assessment but the outcome itself. The more you want it, the more you will give it your all to *make* your company work, to shape the world in the image of your desire.

I'm not a fan of Nietzsche. He strikes me as a deeply disturbed guy who despised humanity. But he was as brilliant as he was troubled. Nietzsche's concept of the *will to power* describes the fundamental driving force behind human endeavors; it influences our predictions by making them work to satisfy our desires.

Individuals are motivated by a craving to expand their power, overcome obstacles, and impose their will on the world around them. The will to power manifests itself in various forms, from physical dominance to creative expression and intellectual prowess. It is the essence of life, the means through which we overcome constraints to become our fullest selves. It can be even more pressing than our desire for happiness; it's what makes us crave stardom, even when we know that fame leads to misery.

Your will to power is your lust for life. It's your thirst for survival, desire for pleasure, and hunger for status. It's the heat of sexual attraction. It's the rage you feel when someone cuts you off on the road. It's the determination to have things go your way. It's the adrenaline of the hunt. It's the appetite for more.

Sometimes, our main desire is not to dominate or grow but to make our pain stop, or to avoid future pain. Sometimes we predict that things will turn out alright because we can't bear the possibility that they won't, and sometimes we predict that things will be worse than they will be as a precaution.

The desire of weather companies to avoid the anger of the public pushes them to increase the probabilities of rain.[2] Most of us will not take our umbrella if we see a 10 percent chance of rain, and when it does rain, we curse the weather app for betraying us, so apps inflate the chances of rain to nudge us into taking that umbrella. Similarly, when storms approach, responsible authorities tend to overreact, because the bad consequences of overreacting are less bad than those of underreacting. In March 2017, the National Weather Service predicted a major blizzard on the East Coast of America. Thousands of flights were canceled. But when the storm hit New York City, less than seven inches of snow fell.[3]

You might think that now that artificial intelligence does much of the predicting, we can be free from people's will to power. But algorithms resemble their creators the way dogs resemble their owners. Algorithms may seem more objective, but that just makes them more dangerous and deceptive. When people at Instagram designed tools to predict which posts would become viral, they craved user engagement so badly that they were willing to get it at seemingly any price, including the mental health of teenagers.[4]

Algorithms are often the hit men sent by companies to carry out their will to power.

Third, *predictions are about power.*

Power is having the ability to motivate someone to think or do something that they wouldn't otherwise.[5] It is also having the capacity to carry out your own will despite resistance, as the sociologist Max Weber put it.[6] To have power is to be able to influence others and exercise force unto them—to make people do, and to do upon them.

Predictions are power moves much more than they are attempts at acquiring knowledge. When the CEO of a tech company predicts that in the future AI will do everything, for everyone, everywhere, he is doing marketing; he's influencing us to want to buy AI. When the great boxer Muhammad Ali, in a press conference in 1971, predicted that he would beat Joe Frazier, he wasn't running probability numbers; he was intimidating his opponent. Predictions wield power through words (they are speech acts, in philosophical lingo).

More often than not, predictions are *meant* to be power plays. And their disguise as a pursuit of truth—or, worse, mere facts—is the perfect alibi. Forecasts look innocent because they resemble facts, and they are related to objective reality (there is a fact of the matter about what the chances are that it will rain). Furthermore, predictions give whoever made the prediction plausible deniability (for example, "I never said it was a fact"; "That it happened doesn't mean I was wrong; I said it was unlikely, not that it was impossible"; "My prediction is still right; it's just a matter of time before it becomes true"). By cashing out actions in terms of prophecies, perpetrators of crimes can elude responsibility. Hannah Arendt writes about how Nazis explained the extermination of Jewish people not as something they were morally (and criminally) responsible for but as the mere fulfilling of a prophecy (they had made).[7]

Even when they are not meant as power moves, predictions wield power. A prediction about the weather will have no effect on whether it rains, but a prediction about anything to do with human beings is likely to change those whom it probed.

Fourth, *predictions are sometimes impossible.*

An implicit assumption of the naive view is that everything is predictable, and if we can't predict it yet, it's just a matter of time until we can, as though there were merely a scientific barrier to overcome. But some things are unpredictable. Worse, the most earth-shattering events are unforeseeable. And if we're not clear on what isn't predictable, we might lull ourselves into thinking we know what's coming, instead of being mindful that we don't.

Fifth, *predictions can be harmful.*

Predictions change the world because they change people's expectations and behavior. There is a choice in making predictions, and that choice carries consequences for people. A forecast that a person or a country will fail can be the cause of such failure. We have been using prophecies to make hugely important decisions, such as waging war, since before the Oracle of Delphi. But the study of the ethics of prediction is a desert, with few exceptions, mostly related to the task of predicting health outcomes during triage in medical ethics. It's about time we start thinking more seriously not only about what predictions are and what they are not but also about when we should make predictions and when we shouldn't. There are some things that we shouldn't predict, even if we could.

To be blind to the speculative, wishful, powerful, limited, and harmful aspects of prediction is dangerous. It makes us clueless about what a prediction is, how it's used, and the impact it can have. The naive view of prediction benefits those who make prophecies and are not naive about them, while the wise view informs the citizenry and helps level the playing field. Whether you are a curious reader, a concerned citizen, a businessperson, or a policymaker, it's better to be wise than naive about predictions.

METHODS OF PREDICTION

Sometimes we make predictions based on gut feelings. Instinct often tracks truth better than conscious logical thinking. Sometimes we know more than we think we know, and we trust our gut feeling as a kind of heuristic, a shortcut. Experienced chess players can know what the right move is before they have thought it through. People demonstrate remarkable accuracy in making social judgments from brief exposures to others' behavior, a phenomenon researchers call *thin-slice predictions*. Observers watching teachers for just ten seconds in a silent video can predict their teaching effectiveness ratings by actual students who had studied with the teachers for a full semester.[8]

Gut feelings are how our body communicates implicit knowledge to us. But it's harder to rely on intuition in abstract and complex situations. In the past, when our intuitions were not enough, we used shamans, oracles, palmistry, astrology, and divination of various kinds. Although astrology and psychics are still popular, they are no longer the official channels through which important decisions are made.

Today, it's more common to make predictions informed, or misled, by data. If you are thinking about opening a retail business, you are likely buying databases containing demographics about the region where you are setting up shop. A busy street, complementary nearby stores, and little or no competition in the area are all good omens for your business to thrive—in theory. But sometimes the data is missing the most important data point, or context that would change the meaning of that data. It might be that a nuclear reactor will soon be built in the neighborhood that will change everything.

The biggest flops are preceded by the highest expectations. The Edsel was named after Henry Ford's son. When the car went into production in 1957, the company had spent ten years and $250 million developing it. Ford had conducted extensive market research and "depth interviews" to design the perfect car. It estimated sales of up to 400,000 Edsels a year.[9] What the research was missing was that by the time the car debuted, a new trend would come onto the scene: a preference for compact cars over mid-priced ones. The Edsel was discontinued in 1959, having sold slightly more than 100,000 cars in total.[10]

Increasingly, we are making predictions with the help (or hindering) of artificial intelligence, which amounts to using even more data with ever more computing power. Your local hospital is probably using AI to predict busy and quiet times and schedule shifts accordingly. How long you wait in the ER when your child has a fever in the middle of the night might depend on a prediction. Your grocery store is using AI to predict how many bananas it will sell today, and whether it needs extra ice cream and cold drinks because a heat wave is coming. When a prediction is inaccurate, you'll see an empty shelf, overripe produce being discarded, or the price of your favorite chocolate go up.

If you run a company, you might be using algorithmic predictions to determine where to invest in real estate, whom to hire, which customers are thinking about leaving, how to optimize your supply chain, and much more.

Which method we use for prediction matters to establish accuracy and legitimacy. But, for the purposes of the main issues this book is about, all predictive methods share enough to be analogous to one another. If you predict that your opponent will lose as a way of intimidating them, it doesn't much matter if you used a psychic, rolled the dice, or hired a predictive algorithm to do your bidding. It might matter to your audience, of course, who might not be impressed if you use an unpopular method for prophesying. But part of what an analysis of prediction reveals is that predictive methods perform the same social and political functions: They are used to influence the future through changing people's beliefs, expectations, and behavior.

Artificial intelligence might seem awe-inspiring and mysterious to the average person, and it is impressive in many ways, no doubt. But that's how the Oracle of Delphi seemed to ancient Greeks, and astrology to medieval Europeans. Although it might be impossible to fathom the inner workings of artificial intelligence (even for the professionals who design it), it is perfectly possible to unveil its political workings, how it is used by power and for power.

The computer scientists of today play the same role in power and prophecy as the oracles of the ancient world, the astrologers of the Middle Ages, and the social scientists of the nineteenth century, as we will see.

A MAP FOR THE ROAD AHEAD

This is a big-picture kind of book, aspiring to contribute to public debate with one overarching idea: We should be thinking much more carefully about what prediction is, what it's not, and how to use it to forge the world we want to live in. Every chapter adds detail to the main idea but aspires to be a contribution to the literature on its own. The book starts slow, with a historical overview, and builds up toward a bang, or rather, a crunch, rip, or freeze (or your favorite version of how the universe ends).

The first part of the book explores the promises of prediction.

Chapter one is a brief history of prediction, focusing mainly on ancient Greece and Rome. Its contribution lies in its analysis of prophecy in the currency of power. It is easier for the motives of prophets to become transparent once we have stopped having faith in old forecasting methods. Prophets are merchants of prediction.

Chapter two tells the story of how we went from qualitative predictions to quantitative ones. It unveils the philosophical ideas that gave way to those developments, as well as the ones that were born out of them.

Those who are not historically inclined can skip the first two chapters, but you'll be missing out on fortune tellers getting thrown off cliffs, seers bringing about the deaths of emperors, and the astonishing discovery that murders and other seemingly disconnected acts follow a predictable pattern in the form of the normal curve; how is it possible that every year we can roughly predict how many murders there will be if those crimes are committed by people who are not trying to fit a statistical pattern?

Chapter three makes the case that artificial intelligence is the new Oracle of Delphi. The likes of Elon Musk, Sam Altman, and Eric Schmidt sharing drinks with presidents and prime ministers is reminiscent of Nostradamus advising Catherine de' Medici, and Rasputin counseling Nicholas II. This chapter links the social and political functions of past prophets with the tech executives of today.

The second part of the book deals with the pitfalls of prediction.

Chapter four delves into the ways in which predictions are similar to truth but are not quite facts. Generative AIs, I argue, are fortune tellers, as opposed to truth tellers, and the ultimate prediction machine is also the ultimate bullshitter.

Chapter five investigates the relationship between prediction, surveillance, and self-fulfilling prophecies (when our belief in a future outcome shapes our actions in ways that bring about that very outcome). This chapter is the beating heart of the book; its philosophical innovation is to argue that although predictions appear to be descriptive claims, they are in fact veiled prescriptive assertions—they tell us how to act. They are what philosophers call *speech acts*. When we believe a prediction and act in accordance with it, it's akin to obeying an order.

Chapter six probes the unforeseeable. It offers a classification and analysis of the sources of unpredictability, or what I call *predictive troubles:* from data troubles, to social, scientific, coincidental, and ironical troubles.

The third part of the book asks us to rethink the uses of prediction.

Chapter seven is a defense of truth, virtue, and beauty through a criticism of effective altruism and utilitarianism. Effective altruists have become the prophets of tech prophets. I argue that the root of what's wrong with utilitarianism and effective altruism is their reliance on prediction. In ethics, ideals outdo predictions.

Chapter eight defends the importance of designing society in a way that makes it possible to defy your odds. When predictions determine our fate, we lose freedom. Democracy needs uncertainty to thrive. It's only when we don't know the outcome of a future election that we have democracy.

Chapter nine invites us to embrace the unpredictable with curiosity. Spaces of indeterminacy are where creativity, humor, and innovation flourish. To be successful—in life, business, and democracy—we need to get comfortable with uncertainty. Resolving uncertainty as quickly as possible leads us to surrender our power to others who end up deciding our future. The chapter proposes philosophy as an antidote to prophecy. Although Stoicism has gained popularity in tech circles and beyond, Epicureanism has better responses to the challenges posed by prediction, encouraging us to be the authors of our own lives.

PART ONE

THE PROMISE OF PREDICTION

CHAPTER ONE

PROPHETS AND POWER

*A brief history of prediction, or why
astrologers get thrown off cliffs*

In Rome, Tiberius was the first emperor to have had a court astrologer. He had fallen out of favor with his predecessor, Augustus, and was exiled in Rhodes when he met Thrasyllus. The future emperor was in the habit of testing astrologers when he needed advice. If they proved to be wrong, he'd have them thrown off a cliff on his way back home. But Thrasyllus impressed him. First, he predicted that Tiberius would succeed Augustus as emperor. Excellent start. Then came the test question: How did Thrasyllus's own horoscope look that day?

The astrologer looked to the stars and visibly panicked. Whether he had heard of Tiberius's cruel habits, had inferred his patron's intentions from his body language, or had indeed learned them from the stars, Thrasyllus answered that he was in grave peril. Smart man. Tiberius embraced him, congratulated him on his divinatory prowess, and assured him he'd escape danger. From then on, the astrologer was his trusted ally. Thrasyllus had shown Tiberius that he understood that predictions are not about knowledge but power.

Prophecy and power are kindred spirits. They walk through the same hallways, share meals with the same people, and exchange money between themselves. Prophets are powerful both because their services are coveted, and are therefore compensated generously, and because their predictions give them influence. In turn, prophets risk the rage of power because predictions can constitute a challenge to power. A prediction that a battle will be lost can itself threaten victory.

Whoever has power runs a risky business. The powerful enjoy the views from above, but they are also forced to safeguard themselves against those who vie for their place at the top. The mighty have often reached the summit through theft, murder, and other unpalatable ways of stepping over others, leaving a trail of resentment in their path. Being too powerful can be as dangerous as being powerless; tyrants rarely die from old age. Since prophets wield authority, they have historically shared the risk that comes with power: One imprudent prediction could cost them their lives. But that has never stopped them from prophesying and being in high demand.

"His business was dread," wrote Toni Morrison of a psychic. Thomas Hobbes wrote that our urge to look into the future is rooted in "perpetual fear." It's good for self-preservation to have a sense of the terrible things that might come.

What would you be willing to give up to avoid the worst thing that will ever happen to you? I would've appreciated a heads-up that the two men sitting across the aisle would rob at gunpoint all the passengers of the bus I was in. It would've been helpful to know that a flood from the neighbor upstairs, who was on holiday, was going to cause my flat's ceiling to collapse a week before I was due to sell it. And I would've surrendered almost anything for a few warnings that came too late. Wouldn't you like a tip-off to give you a chance at dodging the next wrecking ball? You might be able to escape a car crash, or being the victim of a crime, or getting involved with the wrong person. Maybe with foreknowledge you could save your loved one from an untimely death. Can you *feel* the anxiety that drives you to want to know the future? How much would you be willing to pay for an accurate prediction?

Before coming up with a ballpark number too quickly, bear in mind that seeing the future is not only about fear, but also about hope, and desire. It's good for business. Wouldn't it be useful to know which neighborhood will thrive next time you buy a property? Had you known what the future held, you could've bought stocks early from today's most lucrative companies. If you had invested $1,000 in the graphic chips company Nvidia in 1999, your stocks would've been worth more than $3 million in 2025. If you could find out today what tomorrow will bring, you could buy a painting from an unknown artist who is destined to become the greatest painter of the century. Can you *feel* the lust for knowledge of what's ahead?

Prediction is seductive because the future conveys a competitive advantage to whoever can glimpse it. It's a source of great power. That's why people are willing to pay for it, and why it's a good business. Ancient Greeks were willing to sacrifice a whole sheep for the foreknowledge of Apollo.

This chapter illustrates the power of predictions in practice through a brief, and necessarily incomplete, historical tour, mostly focused on ancient Greece and Rome. The social landscapes of the ancient world might seem far away, but they have left lessons we have yet to learn.

DRINKS AND OMENS IN OXFORD

There is always a good reason to have drinks at Oxford. Wine is sipped in start-of-term dinners, mid-term dinners, end-of-term dinners, graduation parties, and before every formal dinner, in which dons wear black gowns and candles light the dining halls of the colleges that are the soul of the academic capital.

On one such occasion, among the mingling academics and the passing canapés which unfailingly look much better than they taste, I spot one of the most knowledgeable people in Oxford: the Bodleian's librarian and author of *Burning the Books,* Richard Ovenden. When he asks me about the book I'm writing, I look away, trying to evade the question, mumbling something vague about prediction. "Oh, we're just about to open an exhibition on oracles and omens in which we explore the universal need to tame uncertainty," he says. Of course they are. The Bodleian Library always seems to have its finger on the pulse of the world.

A few weeks later, I visit the exhibition and take photos of the oracle bones from the Shang dynasty in China, the Egyptian celestial globe, the almanacs, the tarot cards. The casebooks of astrologer-physicians in sixteenth- and seventeenth-century England suggest, first, that people tend to ask about the same things throughout the centuries: We ask about our health, personal lives, business ventures, and political context. Second, in almost all cases, prediction is a business.

So let's start with how most good investigations into human affairs begin—by following the money.

MERCHANTS OF PREDICTION, OR THE BUSINESS OF PROPHECY

Oracles

Divination is not just good for business; it's a good business in itself. Prophets are merchants of prediction. The Delphic Oracle sat on Mount Parnassus, on the northern coast of the Gulf of Corinth, around ninety miles northwest of Athens, near the port of Crisa. We first hear about the Oracle in the *Odyssey,* although Plutarch, who served as a priest at the sanctuary, is a more informative source.

To get to the Oracle, you had to ascend the Sacred Way to Apollo's temple, your sandaled feet treading worn limestone steps that countless pilgrims had climbed before you. The mountain air grows thinner and sweeter as you rise, carrying the smoked scent of burning laurel leaves and the earthy perfume of wild thyme that grows between the cracks of the ancient stones.

Below are the twin cliffs of the Phaedriades tower, like guardian giants, their red-gold faces catching the morning sun. The sound of trickling water from the Castalian Spring, where you stopped to wash yourself and quench your thirst, mingles with the notes of someone singing Homeric verses in the distance, accompanied by a lyre. As you approach the temple, you pass between rows of magnificent votive offerings—gold-rimmed tripods on which oracles sit to pronounce the future, marble maidens, and bronze warriors gleam in the dappled sunlight filtered through cypresses and olive trees.

The temple's massive columns rise before you, their surfaces still bearing traces of vibrant paint—blues, reds, and golds. Inside, it is dark and cool. The air grows heavy with the smoke of incense and the sweet metallic tang that rises from a deep fissure in the earth below; it's the breath of Python, they say, slain here by Apollo himself. Those who inhale the fumes feel intoxicated. Centuries later, the chemist

Jeffrey Chanton will find traces of ethylene in the earth's chasm that could have been the cause of altered states of consciousness, which fits with Plutarch reporting a sweet smell.[1]

Your heart quickens as you approach the inner chamber where the Pythia, the priestess, dressed in a white robe, sits upon her bronze tripod, wreathed in sacred vapors. Originally, the Pythia was a young virgin girl from a respectable family. But after one Echecrates the Thessalian raped one of these young girls, the Delphians decided that the Pythia would be an elderly woman who was bound to celibacy during her service.[2] The old woman's eyes are wild and distant, seeing beyond the veil of ordinary reality. Your palms sweat as you wait for the words that will echo with double and triple meanings, prophecies that will haunt you long after you descend from Apollo's mountain.

The Oracle was not only an experience to remember. First and foremost, it was a commercial enterprise. In the Homeric *Hymn to Apollo,* the god promised the priests of his new Delphic cult that if they built an oracle, they would never again want for food or comfort. A successful oracle put a place on the map and fed its people.

Each visitor to Delphi had to provide a sheep for sacrifice before consulting the god, from which priests could choose a cut of meat. Early Delphic priests had a reputation for snatching the best cuts before anyone else stood a chance. The skin of the sheep was then sold to tanners for a handsome price.

Consulting the Oracle was a lengthy process. The priestess worked only during days in which Apollo was believed to be present at Delphi to channel his wisdom to the Pythia. There were preliminary purification rituals to go through, as well as sacrifices. The queue could be long, and it could be made longer by a high dignitary cutting in. It could take days or weeks to get a response, which was convenient for local hotels and taverns in charge of housing and feeding the visitors. Entertainment grew around oracles. Much as going to Niagara Falls might surprise the visitor who was expecting a natural oasis and finds something closer to an amusement park, places like Delphi and Dodona, the oldest Hellenic oracle, weren't quite the tranquil spiritual havens that the naive might imagine.

Oracles often sponsored athletic games that attracted competitors and spectators from all over the world. The Pythian Games were held every four years at Delphi. The stadium overflowed with thousands of spectators, their excited shouts echoing off the mountain walls as athletes competed in honor of Apollo. In the theater, poets declaimed their newest works while musicians competed on the kithara. The hippodrome thundered with the pounding of hooves as chariots raced neck and neck, wheels nearly touching, while the crowd roared with each dangerous turn. Laurel wreaths awaited the victors.

As dusk fell, celebrations spilled into the streets; singing, dancing, and impromptu performances blossomed, while the smell of roasting meat and wine filled the air. Everyone joined in the sacred festivities. The union of athletic aptitude, artistic excellence, and divine inspiration made the Pythian Games second only to the Olympics in prestige. It was one big, profitable party.[3]

Freelance Seers

Seers unattached to a particular city could also become quite wealthy—especially if they became celebrities or worked for a king or court. Diviners worked like freelancers and were often also magicians. Magic and divination are close cousins, and for centuries were thought to be the same. The word "sorcery" is derived from the Latin word *sors,* a divinatory lot (reading the future through a throw of objects like dice or coins, or through randomly selecting written symbols like cards). While it was illegal in many parts of the ancient world to harm someone with magic, its use was widely accepted. Before Christianity came along, magicians were just another profession.

But divination and magic also go together for business reasons; they are complementary products, like salty chips that make customers thirsty for beer in a bar. Divination can tell you the future, and magic can change it. What good is it to know that something is coming if you can't alter it? Patrons could sometimes intervene in their destiny without the help of magic, but it was in the interest of the seer to recommend a course of action that would necessitate their further services.

Seers followed diagnosis with a prophylactic treatment, analogous to when tech companies prescribe more tech to deal with the problems that arise from tech. Social media companies prescribe AI to moderate vitriolic content that is partly created through the design of algorithms that prioritize "engaging" content. Dating apps are designing AI "wingmen" to generate conversation prompts in the hope of keeping interactions interesting between users.

While established oracles like the one at Delphi tended to be more prestigious, not everyone had the resources to make the journey there. And not every situation could wait for weeks; there were bets to wager, partners to choose, and schemes to plot.

One reason why oracles like that of Delphi are less closely associated with magic is that they had enough business to thrive with only one product. Apollo could afford to limit his services. Freelance seers needed more tricks to make a living. Conjurers made money by offering solutions that mainstream religion didn't. Magicians provided convenience, much as tech does today.

The symbiosis between prediction and profitable action becomes apparent in spells that encourage the magician to revisit the god for a "better prophecy" if the first one was too grim. Magicians were entrepreneurs, and they needed to offer the services that their clientele asked for. Most people are not shopping for news that leaves them feeling despair. If a seer could bend the future in favor of his clients, then all the better for business.

Healers and Warriors

Two of the most prominent side businesses of prediction are, and have always been, medicine and war. Perhaps it's because life is at stake in both, and the more we fear, the more desperate we become for forewarning and hope.

In ancient cultures, healers were also often priests, shamans, or spiritual leaders who used divination as one of their many tools. Prognosis is the second most important task of ancient and modern medicine, the first being treatment. But adequate treatment is unlikelier if the doctor has no understanding of the causes behind the ailment. Prognosis involves understanding causality. It's about reading the signs of the body and their relation to the environment. The right treatment is in essence a prediction that a particular remedy will heal a particular illness. Ancient doctors made use of dreams, astrology, and other forms of divination to prescribe a treatment, in addition to herbal knowledge and common sense.

Predictions were as important in war as they were in medicine. Good strategy involves good predictions. Soothsayers in ancient Greece were often found on battlefields. In the *Iliad,* Calchas gives prophetic advice but also joins the fight. Iamid Tisamenus won five battles for the Spartans with the help of divination. The seer Aristander advised Alexander the Great through many military undertakings.

Before judging ancient people too quickly for turning to seers and astrologers for medical advice and policymaking, bear in mind that divination was the cutting-edge method of decision-making then. It's controversial to call any kind of ancient practice "science," because it is anachronistic, but divination was nonetheless a pursuit for which people trained and learned techniques, and whose objectives were related to truth as well as power. Perhaps a more precise word is the Greek term *technē,* or practical knowledge.

Health and war are also two areas in which our new prophets, the executives of tech companies, are gaining terrain. In health, Alphabet, Apple, Meta, and Microsoft are supplying data infrastructure services to health-care providers, and are involved in home medical surveillance, electronic health records, predictive measures for infectious diseases, wearables for clinical studies, and more.[4] In the military, OpenAI is working with the U.S. Department of Defense developing "frontier AI capabilities to address critical national security challenges in . . . warfighting." Amazon, Google, Oracle, and Microsoft have contracts worth billions with the Pentagon, and Anthropic has partnered with Palantir to bring its bot, Claude, to the U.S. military.[5] In a ceremony in June 2025, four current and former executives from Meta, OpenAI, and Palantir, dressed in combat gear and boots, swore an oath to defend the United States and were pronounced lieutenant colonels in a new unit, Detachment 201, designed to advise the army on technologies for combat.[6] We haven't diverged too far from our ancestors.

READING THE SIGNS

In *On Divination,* the great Roman orator Cicero wrote, "I know of no people . . . who does not consider that future things are indicated by signs, and that it is possible for certain people to recognize those signs and predict what will happen." For prophecy to work, you have to believe that the world is built in such a way that it leaves clues that allow a glimpse of what's to come. Then you need a method to read the signs of the world.[*]

Whether we call it science, techne, trickery, or a combination of all three, the seer was someone who could read the book of the world, whether it was on the canvas of the body, the battlefield, the starry sky, or our dreams. All were instances of the same ability to read the signs of the larger cosmos through its reflections in smaller details of the mundane.

Among the many methods of divination, reading the entrails of dead animals was essential. Ancient Greeks and Romans thought of organs as writing tablets for gods—a metaphor that was taken a tad too literally by a seer who wrote "victory of the king" with ink onto a freshly procured liver.

The metaphor of reading the cosmos is one that runs throughout the history of prediction, and it makes books especially important. Books, one of the threads weaving the narrative of this story, hold weight for us partly on account of the metaphor they embody.

All divinatory methods are ways of reading the signs in the book of the universe. They were all prone to abuse, and abused they were.

PROPHETIC CON ARTISTS

Wherever there is money, there is power, and wherever there is power, there is abuse of power. Herodotus tells us that the priestesses of the Oracle of Delphi were sometimes accused of accepting bribes in return for delivering convenient political messages. And freelance seers were even more susceptible to corruption.

---

[*] If you're curious to know more about ancient methods of prediction, see the appendix.

—

The marketplace of Abonoteichus buzzed with excitement. It was the 150s, and word had spread through the small Black Sea port that Alexander (not the Great, a much lesser one) had returned home—a man transformed, no longer the humble carpenter's son who had left years earlier. Alexander the Paphlagonian had piercing eyes, and he moved with the confidence of someone touched by divine power.

His first miracle was modest but effective. Walking along the muddy shore one morning, Alexander "discovered" a goose egg. As a crowd gathered, he cracked it open to reveal a tiny, newborn snake. People gasped. This was no ordinary serpent, Alexander announced; this was Glycon, the new incarnation of the healing god Asclepius, come to deliver prophecies directly to the people.

Alexander told the locals that the Sibylline Oracle, which no one had heard about before, and some bronze tablets that he claimed had been discovered at a temple of Apollo, commanded that they set up a cult in Asclepius's honor. The prophet's charm sold the locals on the idea.

In the shrine, behind a curtain, Alexander would receive supplicants seeking to know the future, with the fully grown Glycon wrapped around his shoulders. Charisma goes a long way with prophecies, because for predictions to be powerful, people have to believe them, and there are few tactics as effective at earning people's trust as self-confidence. Having the snake perform the astonishing feat of speaking when delivering predictions also helped.

What the gullible didn't see was the elaborate mask that Alexander had crafted for the snake. When the animal was wrapped around his shoulders, Alexander could discreetly tug on horsehairs that would make the mouth of the mask open and close. Nor did the audience notice, in the penumbra of the temple, the speaking tube made from a crane's windpipe hidden in the wall through which an assistant would provide the divine voice for the snake.

Alexander was a con man. And Lucian, a contemporary writer and satirist, was intent on exposing him. The prophet had tried to kill the writer for investigating his methods. But Alexander's assassination attempt only gave Lucian's investigation new life, and his pen new purpose. Word of advice: Don't mess with a writer.

Those wanting to ask Glycon a question were told to write it on a scroll, roll it up, and seal the scroll by stamping an imprint on hot wax. When the scrolls were returned to their owners, they were still sealed, but the questions had been answered. Could that be magic?

Lucian explains that Alexander would carefully open the wax seal with a hot needle and then close it again in the same way after having written his responses. If that failed, he would make a plaster copy of the seal, which he would use to reseal the scroll.[7]

Alexander kept records of his most important prophetic responses. If time showed that an answer had been misguided, he destroyed that record and rewrote history, coming up with a more adequate response.

Despite Lucian's warnings of fraud, Alexander's cult flourished. Even the imperial aristocracy consulted the Sibylline Oracle. Coins representing Glycon circulated for centuries after Alexander's death. Few want to expose a false prophet for whom they have fallen. Once, Marcus Aurelius himself asked for a prophecy from Alexander. The latter declared that the Roman army would be triumphant over the Marcomanni; they were not.

Bringing down con men can be surprisingly difficult, no matter how much of a sham they might be. Prophecies sell well because there's a market for them. We create con artists by craving to know a future that is unknowable. Alexander, however, has gone down in history as the con man he was. In my book, that makes Lucian the victor.

The first lesson that Lucian teaches us is as true today as it was in his time: Where there are predictions, there are con artists. *Caveat emptor;* buyer beware. Second lesson: If more people had read Lucian, they might've avoided being conned, so pay attention to critics of prophets. Third lesson: Where there are predictions, there is also power.

PARTNERS IN POWER

Prediction cannot be disentangled from power. Seers have long been the advisers of rulers. Leading a country is a difficult, lonely, and stressful job to bear. Emperors, monarchs, and presidents are often faced with impossible decisions that could make or break their lives and their countries. Leaders also tend to be surrounded by greed and envy, making for an environment in which it is hard to know whom to trust.

The Nechung Oracle is the personal and state oracle of the Dalai Lama. It was he who, in 1950, under the pressure of Chinese occupation, resolved that the young Dalai Lama be given full powers two years early. "His time has come," he said. In 1956, the oracle prophesied that the light of the Tibetan leader would "shine in the West."

It was again the oracle that advised the Dalai Lama to return to Tibet after he had been on a diplomatic visit to India. Lukhangwa, the prime minister to the Dalai Lama, worried for his life. "When men become desperate they consult the gods. And when the gods become desperate, they tell lies!" warned Lukhangwa. But the Dalai Lama heeded the advice of the seer, only to flee Tibet once more after having consulted the oracle three times. "Go! Go! Tonight!" the medium said in his trance. Had you been a terrified twenty-four-year-old with your people camping outside your palace vowing to protect you with their lives, and Chinese troops approaching, you might've listened to the oracle too.

Another way in which prophecy and power intertwine is through the use of prediction to lend authority. Prophecies can lend credibility to political decisions.

Many cities have been founded in light of prophecies that justified their location. Ancient Mexicas founded Tenochtitlán, today's Mexico City, in 1325, fulfilling an ancient prophecy that the tribe should settle in the place where they saw an eagle devouring a snake atop a cactus—an image depicted on the Mexican flag.

Rome was founded where Romulus spotted auspicious birds. Al-Mansur chose to move the capital of the caliphate from Damascus and founded Baghdad in accordance with astrology.[8] The founders of Kiev chose its site based on a prophecy. Cuzco was built where a golden staff its founders carried sank into the ground, as instructed by the Sun God Inti. Thebes was founded on the instructions of the Delphic Oracle to follow a cow and build a city where it lay down. The Roman emperor Constantine I chose the site for Constantinople, modern Istanbul, after experiencing a divine vision; an angel that no one else could see showed him where the new walls were to be built.

Predictions don't only legitimize political decisions and cities. They also lend authority to leaders themselves.

Suetonius, who chronicled the lives of the emperors, tells us that Augustus, in exile at the time, visited the house of Theogenes the astrologer in Apollonia. When Augustus disclosed the time of his birth, Theogenes flung himself at his feet, giving his patron such a boost in confidence that Augustus even dared to publish his horoscope, and stamp a coin with the zodiac sign under which he had been born, Capricorn. Given that Augustus was a nobody at the time, this prediction would've been all the more impressive. More likely, the prediction of greatness was fabricated once greatness was achieved or within reach.

Predictions of greatness offered power to the person occupying the throne, but also to the astrologer, who could become a lifelong business partner. Correct predictions validated the ruler (who was seen as having the backing of the stars), the astrologer (who was able to read the signs correctly), and astrology itself (suggesting both that there was a settled destiny and that it could be read off the stars).

For the purposes of the enhancement of power, it doesn't much matter how accuracy is brought about: whether it's through a genuine prediction, one made up after the fact, or whether reality was brought to align with the prediction through violence or a self-fulfilling prophecy.

An astrologer who found himself still alive after the time of death he had calculated for himself hanged himself out of respect.[9] It is a good metaphor for how, through predictive algorithms, we are getting too hung up on prophecies ourselves. But we still have the chance to loosen the noose around our necks.

Capricorn became Augustus's personal badge. With the publishing of his horoscope, astrology became part of official policymaking. A few decades later, Tacitus tells us that most people "find it natural to believe that their lives are predestined from birth, that the science of prophecy is verified by remarkable testimonials, ancient and modern; and that unfulfilled predictions are due merely to ignorant impostors."[10] The idea of fate was part of the accepted assumptions of most of the ancient world.[11]

But using astrology to justify your place in power is risky because others can do the same. That is why, in 11 CE, when Augustus felt life slipping away, he made it illegal to consult astrologers in private, or to ask about anyone's death. Similarly, the emperor Severus concealed part of his horoscope to avoid others calculating the date of his death.

RISKY BUSINESS

Prophecy is a dangerous matter for rulers and diviners alike. One unwise prediction could turn into a death sentence. Making a prophecy could trigger a political chain reaction that spiraled out of control.

The Persian king Astyages dreamed that his daughter peed so much that she flooded all of Asia. His advisers told him his dream meant that he should marry her to a humble man. He did. He then dreamed that his pregnant daughter gave birth to a vine that covered all of Asia. His advisers said that it meant his grandson would take over his kingdom. Astyages attempted to kill the baby but failed. Years later, Astyages met his grown-up grandson Cyrus and, alarmed, consulted with his advisers once again. The seers revised their early interpretation and reassured Astyages that his throne was safe. They were wrong. When Cyrus became king, Astyages murdered the seers who had not seen.

Something worth bearing in mind as we move through history into the present is that uttering predictions has ceased to be dangerous for prophets. That's problematic. Institutions are using algorithms to predict, for instance, whether you can pay back a loan, or whether you will commit a crime. And what happens if they get it wrong? Nothing. Not to them, anyway (you, on the other hand, might get arrested). While I don't advise murdering seers, predictions that cause harm shouldn't be issued without consequences to prophets.

—

In the ancient world, the danger with gaining a reputation as a sooth-sayer was that you could find yourself being consulted by a king, and kings can be dangerous creatures.

Aristander of Telmessus was Alexander the Great's favorite seer, but even he was not safe from the rage of his king. When Alexander's army had reached the river Tanais, Alexander ordered a sacrifice to appease the gods before crossing the river. Aristander reported bad omens. Perhaps afraid to cross the river himself, Aristander informed a top commander, Erigyius, of the bad augury. Erigyius went on to tell others. When Alexander learned of the seer's indiscretion, he summoned him, furious. The historian Curtius writes that "the blood ran out of Aristander's face."[12]

You don't have to be a prophet to see that Aristander was risking his job and his life. He quickly changed tactics, assuring Alexander and the rest of the army that the omens were more positive than he had first thought. The crossing would be successful; it was only a matter of being careful.

This episode shows not only how dangerous it was to be a diviner but also how dangerous prophecies could be. The leader of an army could not afford to have his troops fear a bad omen. It's not surprising that Alexander had a complicated relationship with divination, sometimes flirting with the idea of doing without it and sometimes outright defying it.

When Alexander returned to Babylon, astrologers told him to enter the city while facing west. The king ignored their warning, and died days later, possibly from typhoid fever or malaria, though other hypotheses include acute pancreatitis or even poisoning.

Being an astrologer remained a dangerous line of work through the Middle Ages. The court astrologer occupied the summit of the profession. The most famous one was Michael Scot, who studied at Oxford in the twelfth century and ended up working for Frederick II. Legend has it that he foresaw that he would die from the impact of a stone on his head, so he wore an iron helmet that he removed only in church, where he was fatally wounded in exactly the way he had prophesied.[13]

Louis XI, who ruled in the fifteenth century, also kept an astrologer in court. One day, the soothsayer predicted that a lady of the court would die within a week. When she did, Louis was rattled. Either the seer had murdered the woman to prove his accuracy, or he was so prescient that his abilities could threaten Louis himself. The seer had to be murdered. The king ordered his servants that upon his signal they were to throw the astrologer out the window, sending him to a certain death hundreds of feet below. (Some advice to seers: Keep away from precipices and kings.)

When the astrologer arrived to meet Louis, the king asked him one last question before giving the signal: "You are known to have powerful prophetic abilities. Tell me, then, what will your fate be and how long do you have to live?" "I will die just three days before Your Majesty," said the astrologer. That's even smarter than Thrasyllus's response; it included life insurance. Louis never gave the signal.

The astrologer's prediction was not an honest pursuit of truth. It was a clever power play. If there is only one lesson you retain from this book, let it be this: Predictions are often power moves disguised as quests for knowledge. Even when predictions are sincere, they can create currents of political influence that change the world. They *become* power moves.

Predictions were so dangerous in ancient Rome that they became an easy way to get rid of someone uncomfortable. When Balbilus the astrologer suggested to Nero that he could avoid the danger presaged by a comet by murdering a few prominent citizens, he must've known that he was pronouncing a death sentence.

All a soothsayer had to do to cut your life short was proclaim that you possessed an imperial horoscope—a chart that forecast your rise to power. It wasn't unreasonable for a ruler to fear those for whom astrologers had predicted a future in power, even if they didn't believe in fate or in the ability of astrologers. A bad augury could be enough to embolden the enemies of a ruler and weaken his status in the eyes of the people. Such a prophecy carried momentum to make itself come true. An imperial chart therefore made you a target for assassination, and you didn't have much time to turn things around in your favor—to kill before you got killed.

REGULATING PREDICTIONS

The danger of prophecies pushed Rome to outlaw some of them. Prediction of the time and manner of death was illegal for various periods in ancient Rome. It was nonetheless one of the most important activities of the ancient astrologer.

Valerius Maximus, a Roman historian writing around 31 CE, recounts the first documented expulsion of astrologers from Rome in 139 BCE, at a time of unrest. Rulers feared the power of astrologers to manipulate the public. In the years between the deaths of Julius Caesar and Marcus Aurelius, between eight and thirteen decrees were issued to expel astrologers from Rome. Vespasian and Domitian expelled philosophers in addition to astrologers—both dangerous professions for the status quo.

Since being an astrologer was a hazardous line of work, sometimes they joined forces and acted as an anonymous group. In the tempestuous year of 69 CE, astrologers responded to Vitellius's law banning them from Rome by posting their own ruling:

Decreed by all astrologers
In blessing on our State
Vitellius will be no more
On the appointed date.[14]

The conflict escalated. Vitellius executed all astrologers who crossed his path, but even then he didn't manage to prove them wrong. He was executed too, three months later.

THE WANING OF ASTROLOGY

As Christianity gained its foothold in becoming orthodoxy, astrologers felt the grip of the law tighten against them. Divination was seen as a threat to the authority of the Church, and astrology was sometimes considered the invention of demons or fallen angels.

In the *Constitutions of the Apostles*—a collection of early Christian books offering authoritative prescriptions—astrologers, debauchers, magicians, and philosophers are refused baptism. In 358, magic and divination became one of the five major crimes punishable by death. The emperor Constantius warned his entourage that no soothsayer or astrologer would escape punishment or torture.[15] In 409, the emperors Honorius and Theodosius made astrologers burn their books in front of bishops, or face exile. In 425, astrologers and other heretics were cast away.

The message was clear: Looking into the future was the purview of the Christian god; for everyone else, seeking knowledge about what's to come was banned. Christians even came up with a name to make it a sin: *curiositas divinandi*.[16] Curiosity was seen as the desire to know that which should remain unknown. John Chrysostom wrote, "In the same way as murder and adultery are sins and forbidden actions, so also trust in astrology and belief in Fate are perverse, forbidden. . . . In truth, no doctrine is so depraved and bordering on incurable madness as the doctrine of Fate and astrology."[17] In 357, Constantius declared, "The inquisitiveness for divination shall cease forever."[18] What an optimist.

Divination proved hard to root out. It runs deep in human beings. That it was at the heart of Christianity didn't help. The tale of the Magi was one of astrologers who followed a star that presaged the birth of a great king. Much like Augustus using astrology to legitimize his rise to power, stories about the star of Bethlehem were told to legitimize Christ's place.

In Genesis, God says, "Let there be lights in the vault of the sky . . . and let them serve as signs." Thinkers like Origen argued that the stars were not agents, but rather a kind of dynamic writing in the sky, traced by God's hand, and available to divine powers such as angels to read. He ended up being considered a heretic.

Most of the criticisms against astrology didn't deny the sway of the stars in our lives, but they cast a critical light on whether astrologers were able to access those strings of influence. Favorinus, for example, pointed out that not all stars are known, and therefore not all influences can be understood. Pliny the Elder argued that the divine is inscrutable. He also reminded readers that census records show that people's times of birth aren't correlated with their longevity.[19]

About a hundred years later, the mathematician and astronomer Ptolemy would answer this criticism by arguing that only some things are subject to fate, while others are subject to chance. Although

Ptolemy is often remembered as a proto-scientist for his geocentric model of the solar system, his defense of astrology was an important cause for the popularity of his work. Ptolemy has also been accused of fraud.[20] It seems that he fabricated the data to fit his theories, an old trick still practiced today.

From the middle of the thirteenth century, confidence in astrology started waning, even if Christianity never succeeded in quashing it entirely. Fearing the power of astrologers, in 1631, Pope Urban VIII issued a bull banning the practice of astrology for all members of the Church—a prohibition still in force today. The prediction of the deaths of popes and their family members up to the third degree was an offense punishable by death. Through prohibitions, Christians forced seers to keep a low profile, and critics were successful in raising skepticism about divinatory practices. The Universities of Paris and Oxford condemned astrological theories in the 1270s, and astrology went steeply downhill from then on.

Despite its decline, astrology has remained part of the private and public spheres, sometimes still associated with power, leaders, and celebrities. Nancy Reagan's reliance on astrology was revealed on the front page of *Time* magazine in May 1988. Just like ancient Roman emperors, President Ronald Reagan was supposedly converted to astrology by a dramatically successful prediction. The astrologer Joan Quigley warned that late March 1981 would be dangerous. On March 30, an assassination attempt wounded the president, hooking him to astrology. Rumors have it that Reagan selected the hour for the signing of a treaty with the Soviet Union after Quigley studied the relevant horoscopes.

François Mitterrand, the president of France from 1981 to 1995, and Jacques Chirac, his successor until 2007, consulted the astrologer Élizabeth Teissier.[21] Princess Diana regularly consulted her astrologer, Debbie Frank. Diana predicted that she would die in a staged accident; she'd heard it from "reliable sources."[22] Frank, however, didn't foresee her death. "Astrologers can't predict death. So I couldn't tell her not to get into a car, or to stay at home. Some things are tragic fate," she said.[23]

The *Sunday Express* published the first horoscope in the world for the birth of Princess Margaret in 1930, and most newspapers still have daily horoscopes. A quick tour through online stores like Amazon shows that astrology books and tarot kits sell incredibly well. Mediums and psychics are easy to find in cities like New York and London.

In a 2022 survey by the pollster YouGov, 37 percent of Americans between the ages of eighteen and twenty-nine said they believed in astrology. Google searches for "tarot" in the United States have shot up since 2020. Evan Nathaniel Grim, an astrologer who makes forecasts about the news, has 400,000 followers on social media. He also does private readings, charging $750 per session. Allied Market Research, a firm of analysts, estimates that the value of the market in astrology worldwide was $12.8 billion in 2022. Presumably because they are diviners themselves, they predict the market will reach $22.8 billion by 2031.[24]

Notwithstanding astrology's resilience, it has been expelled from the official halls of power. We went on to develop new methods of prediction.

THREADS GATHERED

Predictions gain their power from our fears and hopes. The desire to avoid bad things and the pursuit of health, wealth, and success tempt us to want to know the future, giving merchants of prediction a commercial opportunity. The Oracle of Delphi was a formidable business, and so were freelance soothsayers, seeing healers and warriors.

Prophecies depend on the belief, first, that the world is legible, a canvas onto which gods write out their plans, and, second, that we are able to read the signs of the world. Those beliefs get exploited by con artists who can smell money and fame.

Prediction and power go hand in hand, with prophets having a long history of whispering into the ears of rulers. Prophecies lend credibility to political decisions and leaders. But being so close to power is risky.

Predictions were thought to be so dangerous by rulers that they were regulated in ancient Rome. As Christianity gained more power, astrology and other kinds of divination became unofficial and outlawed practices.

The main promise of prediction is power, and political leaders are not known for their disinterest in power. Plato's ideal state in which rulers are those who are least eager to rule is nowhere to be seen.[25] Few creatures fear as intensely and desire as savagely as rulers—hence why it's so common to find a prophet by their side.

While our methods of prediction have changed, the place of prediction in the halls of power hasn't. The astrologers were replaced first by mathematicians and social scientists and more recently by computer scientists and tech executives. We will visit those two phases in turn in the following two chapters.

CHAPTER TWO

CONQUERING UNCERTAINTY AND TAMING RISK

The mathematization of decision-making

This chapter tells the story of how we went from a world in which each event was singular and mysterious to one in which everything can be categorized, counted, and foreseen. It is a story not only of what we gained—order, scalability, and predictability—but also of what we lost along the way (and what we killed).

Acknowledging that some benefits came at a cost is important if we are to learn from the past. Technology can always be designed differently, and a better understanding of what we've lost can inspire us to design the future in ways that minimize further losses.

As Anil Seth and I walk toward the city center, a familiar metal sound of clickety-clack approaches. It's Sir Roger Penrose, now in his nineties, with his cane. "That was *very* Oxford," Anil says, smiling and shaking his head as the Nobel laureate walks away. I nod, while the thought crosses my mind that the prize has ceased to mean as much as it did, at least to me. Part of it is on account of the flagrant sexism. Marie Curie was only presented with the Physics Prize after her husband, Pierre Curie, had the integrity to refuse to accept the award unless she was included, and things haven't improved much since.[1] But the Nobel has also lost some of its currency through its relying on prediction.

When Barack Obama was given the Peace Prize, many commented on how his achievements for peace seemed thin with respect to previous laureates like Desmond Tutu. Obama himself humbly admitted that he accepted the prize not as a reward for accomplishments but as a "call to action." Arguably, the Nobel Prizes for the advancement of AI in 2024 relied on prediction too. The prize is supposed to go to "those who, during the preceding year, shall have conferred the greatest benefit to humankind." When I give talks to tech-savvy audiences, I ask how many of them think that AI has done more harm than good; it's usually at least half the room.

The computer scientist Timnit Gebru commented on how it seemed that the Nobel didn't "want to be left out of the AI hype train." "I thought they supposedly waited . . . to see the impact," she wrote, noting that AI hasn't yet been responsible for any lifesaving advancements.[2] It is, however, consuming vast amounts of natural resources.[3] *The Economist* recently called AI "the biggest gamble in business history."[4] Betting on AI is itself a prediction that the ultimate prediction machine will lead somewhere good.

As a neuroscientist, Seth is the first to recognize how deep prediction runs in us. When he takes the stage at the Oxford Playhouse to give his Simonyi Lecture, he explains the theory of predictive processing: Even though it might seem as if we were perceiving the world directly, that is only an illusion. What our brains are doing instead is a kind of controlled hallucination; controlled because it's limited by how the world in fact is. Our expectations, based on our experiences, constantly shape the content of our perception. Our brain makes a prediction that molds our perception, but if feedback from the world suggests that we got things wrong, then we adjust our prediction and perception.[5]

You might see what Seth means when you're expecting to see a friend at a coffee shop and you mistake them for someone who vaguely looks like them. Or when you're thinking about someone intensely, perhaps because you miss them, and strangers seem to take their shape. Your expectations and desires influence your perception. When you make a mistake playing the piano, you might wince as your expectation shatters with the sound of the wrong note—valuable visceral feedback to adjust your perception and tighten the synchrony between the sheet in front of you and your fingers.

In *On Divination,* Cicero distinguished between two kinds of divination, the natural and the artificial. The first is spontaneous, intuitive, and inspired, while the second is inductive and inferential. If Seth is right that we perceive, think, and feel through "natural" predictions, that may be part of why we find "artificial" predictions so compelling. In a loose way, they mimic how we make sense of our experience.

In the times of Cicero, technical, artificial, or inductive predictive methods included the reading of entrails, birds' flights, and the stars. Today we use statistical models—especially AI. But between reading the stars and AI there was a process of quantification that helps explain how we think about and use predictions today.

There are a few main investigative leads to follow in any human story and metaphorical murder mystery: ideas, people, tools, and motives.

THE IDEAS

Philosophical ideas usually underlie most great changes in history. That's why philosophers have sometimes ended up being sentenced to death, exiled, or silenced. Those who believe philosophy is an abstract and useless pursuit are being duped by people who are hiding their philosophical underpinnings. Having philosophical views is inevitable. But some views are more rigorous than others; you can be more or less aware of what philosophy you are subscribing to and its implications; and you can be frank about your philosophical convictions, or you can hide them.

There is everything at stake in our philosophical views, from how resilient we are as individuals to the health of our societies and our very survival as a species. As the philosophy professor Luciano Espinosa once said to me, "Nos va la vida en ello." Our very lives depend on our philosophical views.

There were at least three philosophical ideas underlying the mathematization of prediction. First, a disenchantment of the world that made it conceivable that fate wasn't sealed by gods. Second, an idea of the world as a complex mechanism that can be read through the signs it gives, which led people to think of nature as a book and of signs as evidence. Third, a newfound trust in numbers as the objective language of nature, as public knowledge that could replace private judgment and promote confidence in policymakers.

Disenchantment

Probability, as we understand it, didn't exist in the ancient world. Ancient Greece and Rome, as well as the ancient Arab and Asian worlds, developed impressive technological and theoretical knowledge, from being able to construct buildings that we are incapable of imitating today, to developing refined philosophical theories. But none of them had a mathematical concept of probability.

Some Greek thinking hinted at notions related to probability. Aristotle thought that events were either certain, impossible, or contingent (possible but not necessary), which reveals a concept of the counterfactual—what could have been but was not. Greeks recognized that more things could happen in the future than *will* happen. Plato called the natural sciences "the science of the probable," as opposed to more theoretical sciences of the necessary. The Greek word *eikōs* meant plausible or probable, but such ideas were never translated into mathematics or methodical decision-making.

Even the Arabs, who masterfully transformed Hindu numbers into mathematics and measurement in astronomy, navigation, and commerce, did not develop a theory of probability. Al-Khwarizmi, a mathematician who lived around 825, wrote the earliest documented work in Arabic arithmetic. It's where we get our word "algorithm" from, meaning rules for computing. Al-Khwarizmi inaugurated our rules for adding, subtracting, multiplying, and dividing.

Perhaps what stopped ancient cultures from developing probability was their worldview, their philosophical commitments. If the future is ruled by gods and fate, then calculating the mathematical chance of something happening is nonsensical, unthinkable. If God or fate decides what becomes of you, the only possibilities are to wait and see or try to read the signs in the sky for a heads-up. Math had nothing to do with fate. What happened for probability to develop?

Max Weber's concept of "the disenchantment of the world" (*Entzauberung der Welt*) describes the cultural process in Western society whereby magical, mystical, and spiritual ways of understanding reality were gradually displaced by rational, scientific, and technical explanations.

The rise of Christianity in the Western world brought about a significant shift in worldviews, including a partial disenchantment of the world, or perhaps more precisely, the first step toward disenchantment. The process involved the gradual suppression and discrediting of preexisting pagan practices, including astrology, divination, and nature worship. As Christian doctrine became more dominant, it emphasized a monotheistic perspective that relegated many supernatural elements to the realm of superstition or heresy. Christianity killed all magic except the one attributed to God; it was only natural for people to eventually feel skeptical about that too.

In the fourteenth century, a group of Oxford scholars nicknamed the Oxford Calculators developed and amended Aristotle's thoughts on physics by quantifying the movement of an object through time and space.[6]

Even art surrendered to the allure of measurement. "Measured music," with novel and more precise notation, was the product of and gave rise to more complex arrangements, getting banned by Pope John XXII for "intoxicating rather than soothing the ear." Painting went from a portrayal of reality as understood by the mind (not the eye), in which it was possible to show the inside and outside of a building at once, to Renaissance men like Filippo Brunelleschi and Leon Battista Alberti bringing perspective into representations.[7]

Illustration by Albrecht Dürer of a medieval artist using a quantified grid to capture perspective

The Protestant Reformation in the sixteenth century, and later the Enlightenment of the eighteenth century, led by philosophers like Voltaire and Kant, continued the process of disenchantment, further emphasizing reason and empiricism over religious authority, laying the groundwork for modern secular society.

God and the supernatural got progressively sidelined. Kant, for instance, pointed out that it can never follow from God's commands that we should abide by them. We need another premise that we should always do what God commands. The challenge is that we cannot know that we ought to obey God, that his commands are worthy of being followed, unless we have an independent moral judgment through which we can assess God. But if we possess such a standard, then God's commandments are redundant. For Kant, rationality is at the center of ethics.

With God gradually moving into a background position, the grip of fate loosened. Along with worldviews that could make sense of and contemplate the notion that chance might play an important role in the world, the invention of printing with movable type in the middle of the fifteenth century facilitated the development of abstract numerical thinking, and with it, scientific inquiry and bureaucratic organization.

Weber argued that this transformation was a defining feature of modernity, driven by the rise of scientific thinking, bureaucratic organization, and Protestant rationalism. While this rationalization brought greater efficiency, predictability, and technological progress, it also led to a loss of meaning and wonder in human experience. Where ancient peoples once saw spirits in trees and divine purpose in natural events, modern individuals increasingly came to view the world as a mechanistic system governed by calculable forces.

This shift created what Weber saw as an "iron cage" of rationality, where technical knowledge trumped questions of value and meaning, leaving modern humans in a disenchanted world that, while more knowable and controllable, was also stripped of its meaning.

A Clockwork Universe

The second philosophical idea that gave life to probability theory was a sense of the universe as a complex mechanism, much like a fine clock. The disenchantment of the world and faith in a clockwork universe go hand in hand, one feeding the other, and both depending on quantification, but they are distinct in that a disenchanted world doesn't have to be mechanical (for example, a quantum world).

The invention of the clock transformed time from a steady flow tied to the sun and bodily rhythms into a quantified unit dictated by a mechanical artifact. But it also gave ideas to philosophers. What if everything in the world functioned not owing to some kind of divine breath but thanks to complex mechanisms? What if the universe itself is a *machina mundi,* a world machine? René Descartes wrote that "there are no rules in mechanics which do not hold good in physics . . . for it is not less natural for a clock . . . to indicate the hours, than for a tree . . . to produce a particular fruit."[8]

To read the clock of the world, one had to speak its language, and that language was mathematics. As Galileo put it,

> Philosophy is written in this grand book, the universe, which stands continually open to our gaze. But the book cannot be understood unless one first learns to comprehend the language and read the letters in which it is composed. It is written in the language of mathematics.[9]

For these thinkers, the ultimate source of knowledge is not in books written by fallible human beings but in the book of the universe. Nicholas of Cusa, in his dialogue *Idiota,* writes about curing ignorance, not through reading books by human beings, but by reading the books of God, which "he wrote with his own fingers" and can be found "everywhere."[10]

In the early Renaissance, books were revered as authorities. Fittingly, until then the word "probable" meant "approved by the wise, or by authority." In early modern Europe, intellectuals went from reading books to reading nature. In the early Renaissance, people providing the evidence of testimony and expertise ruled supreme. In the Enlightenment, the evidence provided by things ruled supreme.[11] In this vein, the physician Paracelsus boasted of not following the textbooks of Hippocrates or Galen but basing his profession "upon experience."[12]

In the seventeenth century, experimental forms of reasoning became the gold standard of inquiry. In some ways, this notion was not dissimilar from ancient ideas of reading the signs of the world. Just as the physician read his patient from signs such as her skin, eyes, or urine, so scientists intended to read the atomic world through diagnostic tools. Except in the seventeenth century signs were being interpreted not as messages from gods, but as evidence.[13]

The belief that the world is nothing more than a complicated clock and that we can access its mechanisms is reassuring. It suggests that nothing is random, and everything is knowable, at least in principle. It paved the way for determinism.

The French mathematician and astronomer Pierre-Simon Laplace, whom we will meet again soon, starts his *Philosophical Essay on Probabilities* by stating, "All events, even those which on account of their insignificance do not seem to follow the great laws of nature, are a result of it just as necessarily as the revolutions of the sun." All events are inexorably caused by previous events, all governed by laws of nature.

There were some dissenting voices. Xavier Bichat, a French pathologist known for his view that the human body has different kinds of tissues, taught his students that things are either organic or inorganic, their properties vital or non-vital. Laplace was right about the physical sciences that deal with the inorganic and are subject to invariable laws. Physical phenomena can be "foreseen, predicted and calculated." But bodies are tied to life. Vital functions "defy every kind of calculation, for it would be necessary to have as many different rules as there are different cases." Events in an organism are caused, but every cause is unique, much like its effect, argued the vitalists.[14]

Vitalism lost that battle, and physics became the apex of the sciences, dominating ways of thinking in other fields, from biology to policymaking. But physics turned out to be chancier than philosophers thought. In the twentieth century, chance smashed determinism.

With a belief in a clockwork universe, we weakened the authority of experts, and we killed the idea of a living world that couldn't be reduced to a mechanism.

In Numbers We Trust

Trust in numbers was partly born out of distrust in people.

It was not only the Church's disapproval that led to a decline in traditional forms of divination; common people were tired of depending on experts to catch a glimpse of the future. Reading animal entrails or the flight of birds is not obvious, and astrology is technical. The opaqueness of these methods facilitated fabrication, errors, and corruption. Divinatory lots such as dice are ways of potentially communicating with the gods in an open way, visible and readable to all. They also invite asking questions about randomness and probability, which might have contributed to probabilistic thinking. But dice were unfit for purpose in the Enlightenment and beyond. Games of chance were associated with gambling and vice, and leaving decisions to pure chance was counter to rationality. Statistics came to the rescue.

After getting a kick start from astronomers, statistics developed mostly thanks to bureaucrats who were responding to challenges in public life. It was a practical endeavor, not a scientific one. Merchants, bankers, farmers, militaries, and tax officials needed to count in standardized ways. Extensive trade needs prices and measures, and therefore extensive quantification. As bureaucracies grew, standardization and public justification were necessary. There was also the question of how to hide power struggles to make them look as rational as possible. Numbers, standards, and statistics make politics and power invisible. That's convenient to anyone wanting to justify what they do as free from values.

Behind our trust in numbers is a closely related philosophical idea: that values cannot be derived from facts. The philosopher David Hume argued that moral statements ("ought" claims, those that tell you what to do) cannot be logically deduced from purely descriptive "is" statements. For Hume, from the fact that there is a child in need and you could save her, it doesn't follow that you ought to save her. Empirical facts alone are insufficient to establish moral conclusions.

The fact-value distinction makes numbers appear as facts free from values. If we can decide public policy on the basis of numbers, we can pretend to leave politics and partisanship aside. If only numbers were so clean. We trust numbers because we don't trust people, but it's people who come up with the numbers.

Numbers are so compelling to human beings that when we focus on them, we can lose sight of other, more important aspects of reality. Johann Bernoulli was part of a well-known family of Swiss mathematicians in the eighteenth century. Bernoulli was so obsessed with numbers that when he went into a room with paintings, instead of focusing on the strokes, the colors, or the scene depicted, he would take out his yardstick and measure their dimensions.[15] It's a good example of how human beings are capable of entirely missing the point when we focus too much on measurements.

We killed trust in the judgment of experts and authorities and replaced it with trust in numbers. And although numbers have been beneficial for science and business, by focusing too much on them, we lost sight of the unquantifiable.

THE SUSPECTS

People are idiosyncratic creatures. We are moved by our backgrounds, personalities, and curiosity. Most of the characters who contributed significantly to our understanding of chances were scholars, money-makers, or bureaucrats. This section is not only about particular people who had an influence on the theory of probability but also about how probability changed how we think about people.

The Gambler

Girolamo Cardano wrote about syphilis, innovated in the way surgeries were done, and was a pioneer in hygiene, recommending showers and baths. When someone leaves a soapy scent in their path, I think of Cardano with gratitude. Cardano is one of the characters of this story who was both a polymath and keen on not losing his money. He was the most famous physician of his time, being known from Italy through Paris and Holland all the way to Scotland, where he was brought at great expense by the Duke of Hamilton.

Cardano was also an astrologer, astronomer, philosopher, and chemist. Apparently, when people didn't have to spend half their time responding to emails, they could do more than one thing in life. He also happened to be a gambling addict, which led him to write the first analysis we have of games of chance in 1564 (although it wasn't published until 1663), along with hundreds of other works in science.

He recognized that the probability of an outcome is the ratio of favorable outcomes to the total possibilities or opportunity set, or, more simply, the frequency of a particular event in a repeated experiment under the same circumstances. This is a *frequentist* definition of probability; it sees probability as a property of the world. Cardano was the first to figure out that in the game of dice the number 7 came up most often, and middle numbers like 6 and 8 appeared more frequently than extreme ones like 2 and 12. And he realized why that was: While 7 can be rolled in six ways with two dice, 2 and 12 can come up only one way (two ones, and two sixes).

Cardano killed the idea that dice were completely unpredictable and inscrutable. The result of any one cast of dice is unpredictable, but patterns emerge with enough throws. Probabilistic ideas were ripening, but it wasn't until the latter half of the seventeenth century that they bloomed, simultaneously and in some instances somewhat independently, in different fields.

The Mathematician

Abraham de Moivre discovered the structure of the normal distribution and the concept of standard deviation through flipping a coin, again and again and again.

If you want a child to discover the normal curve for themselves, flip a coin ten times and record the number of times it lands on heads. Then repeat nine times for a total of one hundred flip coins in batches of ten. You'll see that in most batches a coin will land on heads four, five, or six times out of ten, but now and again you'll get a run of one or two heads, or eight or nine. The more flip coins you do, the clearer a normal curve you'll see. A thousand flip coins will make for a very tidy curve. There are worse ways to spend your time.

De Moivre was a French mathematician living in London. He was friends with Edmond Halley and Isaac Newton, and a member of the Royal Society. He belongs firmly in the category of scholars; he never applied his brilliant mathematical insights to profit making and was always poor. In 1711, he wrote a book about probability in which he defines risk for the first time as the chance of loss. The true measurement of risk, he writes, is "the product of the Sum adventured multiplied by the Probability of the Loss."[16]

In his old age, de Moivre noticed that he was sleeping about fifteen minutes longer each night. Being a mathematician, he calculated that he would die on the day when his sleep reached twenty-four hours, on November 27, 1754. And on that very day he died, as predicted. Unfortunately, this story is likely apocryphal. It's worth noting it, however, because it shows how gullible we are to those who seem to know more than we do. Magical thinking sprouts in the face of impressive mathematical discoveries and proofs.

The Minister and the Philosopher

Thomas Bayes's year of birth is fittingly uncertain, but it was around 1701. He was an English Presbyterian minister who laid the foundation for modern methods of statistical inference with one paper. He was considered a Dissenter or Nonconformist for failing to support the Church of England. As a Dissenter, he was barred from English universities; their loss.

According to his biographer, unlike many other characters in this story, Bayes wasn't in it for the money or the power or even the recognition. "He was more of an amateur, a virtuoso. He did it for his own pleasure rather than having a research agenda."[17] That power warps many or even most human undertakings doesn't mean it seduces everyone or everything.

Bayes had a close friendship with Richard Price, a fellow minister and philosopher who discovered, refined, and published Bayes's most significant work posthumously, roughly a decade after Bayes had written the original paper.[18] If it hadn't been for Price, we might've never heard about Bayes.

Price worked in a chapel in London. Among its congregants was Mary Wollstonecraft, the author of *A Vindication of the Rights of Woman* and the mother of Mary Shelley, who wrote *Frankenstein*. Price was also friends with several of the Founding Fathers of the American Revolution, including Thomas Jefferson and Benjamin Franklin, and with the philosophers David Hume and Adam Smith.[19]

Bayesians, who stand in contrast to frequentists, see probability as an expression of our degree of certainty about something and not as a property out there in the world. The core idea behind Bayesian statistics is that we should update our probabilities based on new evidence, which allows us to incorporate prior knowledge and continuously refine our understanding as more data becomes available. In the language of the statisticians, we compare posterior probability with our priors. We will come back to the debate between frequentists and Bayesians in chapter four.

The Jurist

Nicolas de Condorcet believed in progress. With a quiet outwardly manner and a fiery inner passion, he opposed slavery and the death penalty, and argued in favor of gender equality and universal education. "The time will come when the sun shines only on free human beings," he wrote. Unfortunately, he also imagined a system of universal data collection, capturing every fact imaginable, including moral wisdom—a totalitarian system likely to lead to a dystopia.[20]

Condorcet was interested in juries. He feared that France would copy the English model unreflectively. Why did the English have (and still have) juries of twelve? The case of juries shows how reliance on numbers went beyond science and business to significantly influence how we view moral issues and questions of policy. Centuries later, algorithms are seeping into the justice system, from policing to sentencing.

Condorcet favored a jury of thirty. But, if you had to have a jury of twelve, he argued that the best twelve-person jury would be one that convicted with a majority of ten or more members—the current system in the U.K. If you've ever tried to get ten people to agree on *anything,* let alone an event that they didn't witness, you'll likely have a healthy sense of skepticism toward the justice system.

Part of what is at stake is the kind of society we want to live in. In most liberal democracies, it is common to find people echoing the belief that it is better to let ten guilty people roam free than to imprison one innocent person. It's called Blackstone's ratio, in honor of the English jurist William Blackstone, who first expressed the idea in the 1760s.

But how much do we want to err on the side of freedom? What if it's one hundred guilty persons who go free? A thousand? A hundred thousand? A state in which every single crime committed ends in punishment would be a police state. But one in which no crime gets punished would be a failed state. What is an adequate conviction rate for a healthy rule of law?

The stakes are high. Following Condorcet, Laplace went on to study the same topic: How do we build juries to ensure justice? He made three assumptions: that the probability of guilt of an accused is 50 percent, that the reliability of a juror is between 50 percent and 100 percent (if it was below 50 percent, we'd be better off tossing a coin than having jurors), and that the reliability of jurors is uniformly distributed (that is, it's as likely to be any value in that interval). According to his calculations, a jury of twelve is too safe: We will be failing to convict people who we can be fairly confident are criminals. If we strive for an error rate of one in a thousand, a unanimous jury of nine is appropriate.[21]

But Laplace didn't have any empirical data to contrast his theories with. When the French mathematician Siméon Denis Poisson looked at the matter in light of data, he concluded that the real empirical probability of error was acceptable on a simple majority system. The U.S. and the U.K. haven't caught up. Our systems are strongly biased in favor of the defendant to the point of decriminalizing serious offenses.

Every year, about 400,000 rapes are perpetrated in the U.K. Of those, about 70,000 get reported to the police.[22] Of those, only around 4,000 end up in the Crown Prosecution Service, only about 2,000 stand trial, and only about 1,000 or fewer end up in conviction.[23] That makes the impunity rate for rape around 99.8 percent. By having an almost perfect impunity rate, countries like the U.K. have effectively decriminalized rape.[24] The numbers in the United States aren't flattering either, an expert in the topic tells me on condition of confidentiality; that they are not public is telling enough.

What jurists in the nineteenth century and even now fail to take into account is how much what we want to believe affects our judgment. White guys are our heroes, from Jesus to Superman. We don't like thinking of white men as criminals, so we tend to punish them less often and less harshly. And wasn't it Eve who drove Adam to temptation? To many of us, it is easier and more comforting to believe that a woman lied than that a man raped her. What hides behind seemingly objective numerical choices like the number of jurors are subjective preferences.

The sociologist Donald Black proposed a theory called "relational distance": We punish according to how close or far from the core of a society's tribal identity are the perpetrator and the victim. The killing of animals is often not punished, and killing women is punished less severely than killing men. "I don't know of a single case of a man killing his wife in which the man received capital punishment," said Black.[25] Black's theory partly explains why so many Black men are in jail, why punishment for rape is so rare, and why we choose a certain number of jurors or how we use algorithms in the justice system. The wishful aspect of probability is still as woefully ignored as it was in the nineteenth century.

When Condorcet criticized the new French Constitution of 1793, he was accused of treason. He went into hiding but was discovered, and died during his first night in the village jail, possibly of a suicide pill.[26] He faced no jury, of any number, making him one of the many children of the Revolution who was devoured by it.

The Opportunist

Pierre-Simon Laplace had a taste for power. He was no Isaac Newton, a contemporary scientist who was a recluse obsessed with truth. Laplace's genius wasn't limited to mathematics, astronomy, or philosophy. He was a gifted opportunist who lusted for prominent positions in the public sphere.

When he was twenty-four years old, Laplace was elected to the exclusive Académie Royale des Sciences. He was a founding member of the Bureau des Longitudes, charged with improving nautical navigation and standardizing timekeeping. He was also professor at the École Normale and the École Militaire in Paris, where he examined a sixteen-year-old sublieutenant called Napoleon Bonaparte in 1785.

His abilities as an operator allowed him to stay on top of political turmoil, shifting his allegiance according to convenience, first from the republic to the empire in 1799 and then to the Bourbon monarchy during the restoration of 1814, managing not only to keep his head atop his neck but also to acquire the title of marquis in the process.

When Napoleon was in power, Laplace was appointed minister of the interior, but he lasted only six weeks on the job. Laplace, said Napoleon, "carried the spirit of the infinitesimally small into administration." It wasn't a compliment.

Like many other enthusiasts of counting, Laplace had an authoritarian bent. When the American astronomer Joseph Lovering visited him, he lost some of his respect for Laplace when he overheard his wife asking him for permission to get the key to the sugar cupboard.[27] His contemporaries considered him arrogant. A Swedish astronomer who visited Paris wrote that Laplace was not shy in letting it be known that he thought of himself as the best mathematician in France.[28] And he developed a reputation for using others' mathematical innovations without acknowledging their work.[29]

Laplace's main mathematical insight was that when determining the position of an object in the sky, the mathematical combination of many observations, each subject to random error, leads to a more precise measurement than individual ones. In other words, because it's hard to figure out where a star sits in the night sky, your best bet is to make many observations and average them out.

Laplace combined theory with practice, showing that probability theory could be used to solve problems of many kinds. He used statistics to estimate population sizes, analyze demographics, and calculate the probability of judicial mistakes.

Laplace is responsible for defending determinism. He wounded people's belief in free will. He thought that every event is the inevitable and necessary consequence of previous events combined with the laws of nature. If we had knowledge of all the atoms in the universe and all the laws of nature, then we'd know it all, past, present, and future. But if everything is predictable, then there's no such thing as free will; we are just another billiard ball being pushed around without the freedom to do otherwise.

You might think that there is a tension between probabilistic thinking and determinism. But not so fast. It depends on how we interpret probability. Laplace thought that if we could know all the variables precisely, then we'd know the future just as precisely. But in the real world, we are missing data. Probability theory allows us to make pre-

dictions based on incomplete information. He had a Bayesian interpretation of probability (though he wouldn't have called it that). The
world is deterministic, and we need probability only because we have
incomplete information, thought Laplace.

Is it a coincidence that Laplace was a staunch believer in determinism and had questionable character traits? It's reassuring to think that
you're not responsible for your faults, that necessary causes made you
do it.

The Astronomers

Giuseppe Piazzi, astronomer to the king of the Two Sicilies, was com-
piling a catalog of stars. Gazing into the heavens, he discovered a
new planet between Mars and Jupiter. He named it Ceres, after the
Roman goddess of grain. He was reluctant to announce his discovery
before acquiring further observational confirmation, but he lost track
of the dwarf planet to the glare of the sun. Another scientist would
be able to find it for him.

After coin tossing and games of chance, another route to study the
normal curve came through the observation of the skies. Errors are
inherent to any measurement process, and measuring the distance
between the stars is an exceptionally difficult task.

The stars move, they are obscured by passing clouds, and they are
so far away that a minuscule measurement error from Earth can make
a big difference. If you're out at sea and you make a mistake of just
one degree, it can make you miss the tiny island you were aiming for
in the middle of the Pacific. The challenge was to compare a model
of the heavens with imperfect observations. With the improvement
of tools, especially the telescope, came a greater sensitivity that registered more phenomena with more accuracy but also introduced more
errors (for example, through inconsistencies in lenses).[30] Astronomy
had an advantage: It was the first of the sciences to gather a large
number of roughly accurate data points.[31]

—

Carl Friedrich Gauss was a mathematical prodigy. When he was three years old, he caught a math error his father had made.[32] Gauss had developed a new method of orbit calculation, and in Piazzi's lost planet he saw a chance to test his new theory of errors. Taking Piazzi's scanty observations, which covered only three degrees of Ceres's orbit, Gauss predicted the time and place of its reappearance in the night sky, almost a year after it had vanished into the heavens. Thanks to Gauss, the Austrian astronomer Franz von Zach found the misplaced planet, exactly where Gauss forecast it would be.[33]

How did Gauss do it? He figured out that as long as there isn't any systematic error (for example, an imprecise instrument), the distribution of random errors follows a normal curve. That's why averaging multiple measurements gives a better estimate than any single one: Most of our results tend to cluster around the true value, with more pronounced mistakes dwindling in number the further away we stray.

That is, Gauss took Laplace's insight that an average of many observations is better than one and took it to the next level by realizing that random errors follow a pattern that could be counted on for the purposes of prediction. That insight revolutionized scientific practice, and it's why the curve is also called *Gaussian,* and *the law of error*. It's a kind of "mathematical alchemy," as James Vincent puts it, "transmuting error into accuracy."[34]

The bell-shaped curve has been crucial for mathematical prediction. It allows us to know that for those phenomena amenable to a normal distribution, 68 percent of the observations will fall within one standard deviation of the mean, and 95 percent of them will fall within two standard deviations.

A day after Gauss died from a heart attack, his brain was removed and studied by Rudolf Wagner, who found its mass to be slightly above average, at 3.29 pounds. In 2013, scientists at the Max Planck Institute revisited Gauss's brain and concluded there was nothing abnormal about it after all; it was just another error of measurement. They also realized that Gauss's brain had been mixed up due to mislabeling with that of the physician Conrad Heinrich Fuchs shortly after the first investigations. Both brains are "anatomically unremarkable."[35]

The First Social Scientist

Adolphe Quetelet found it astonishing that we can know in advance roughly how many people will "dirty their hands with the blood of others" every year. Born in Ghent in 1796, he earned the first doctorate ever awarded in mathematics by the University of Ghent when he was twenty-three years old. According to his own account, he was the first person to realize that criminal statistics suggested invariable patterns. When Parisian judicial statistics were published, he noted "the terrifying exactitude with which crimes reproduce themselves."[36] He named his discipline *social physics*.

In 1844, he argued that many human phenomena follow the same curve or distribution that has been associated with coin tossing. This intellectual development is the direct precedent of our algorithmic times. By applying the same mathematical curve to biological and social phenomena, he changed how human beings would think about the mean. He turned the mean from a mathematical abstraction into a tangible character.

In the early 1830s, Quetelet wrote about the *homme type,* "the average man," culminating in his 1835 *Treatise on Man.* Alluding to but misinterpreting Aristotle, Quetelet wrote that the average man's faculties were "in a just state of equilibrium, in a perfect harmony, equally distant from excesses and defects of every kind, in such a way that . . . one must consider him as the type of all that is beautiful and all that is good."[37] (Unfortunately, Aristotle's virtuous person is nowhere near average. Have you noticed how rare bravery is?)

Few liked the concept of the average man except Quetelet. For starters, it was a man, which automatically ignored more than half of the population (alternatively, the "average" person would be neither a man nor a woman). It was also a very mediocre individual. Upon inspection, it is obvious that the average person doesn't exist. And yet the concept stuck, like an annoying jingle that you can't stop singing in your head. One of the many unfortunate effects that the average man had was to reify the concept of a racial type, a people or nation.

Quetelet turned a theory of measuring unknown quantities (stars) with a probable error into a theory of measuring the ideal properties of a population. He turned the statistical from descriptive into normative, from a curious observation about how human beings make mistakes in measuring stars into how we measure and judge our fellow citizens.

What's even more surprising is that Quetelet came to his conclusion with very few examples of normal distributions. Weight, for instance, wasn't typically recorded. Before Quetelet came along and sold us the idea that normal distributions refer to some real characteristic of a population, we had no reason to collect such information.

Quetelet found one example in *The Edinburgh Medical and Surgical Journal,* which in 1817 had published the height and chest measurements of 5,738 Scottish soldiers. He made the following interesting link: If someone not well practiced in the art of measuring the human body was to measure a man 5,738 times, those measurements would be grouped with the same regularity as the 5,738 measurements of Scottish soldiers. Or, in other words, if we received both sets of measurements, we wouldn't be able to tell them apart and know which measured one individual and which measured 5,738.[38]

This idea was so catchy that within weeks people were measuring and plotting every human, animal, and vegetable attribute imaginable. Once Quetelet made mean characteristics into real quantities, the scientific community started thinking of departure from the mean not only as errors (as in the case of measuring stars) but as deviation caused by nature.

Normal People

Are you a normal person?

Our ideas about what is normal have been crucial in the mathematization of decision-making. *Norma* is Latin for a carpenter's square, a right angle. And "right" doesn't only refer to a ninety-degree angle. It has the connotation of being proper or good. If the angles on your door are not right ones, they are the wrong ones. The entrance to the Senior Common Room of Hertford College has a door with wrong angles; it's my favorite door in Oxford.

We use the word "normal" to express both how things are and how they ought to be, and we often do it at the same time. Not stealing is a norm that denotes both what is right and the fact that most people pay for what they consume most of the time. We also use the word to refer to what is mean, or average.

While the great thinkers of the Enlightenment were concerned with understanding human nature, in the nineteenth century scientists became obsessed with understanding normal people, and by association, freaks of nature. Although perhaps it happened the other way around, because statistical regularities were first noticed in connection with "deviancy": suicide, crime, vagrancy, madness, prostitution, and disease.

Auguste Comte invented the word "sociology," but he thought it was so ugly that it would never catch on. He wasn't thinking about the normal as a statistical concept. He thought of the normal as the usual; what was not normal was merely a variation from what was typical.[39] Comte had a soft spot for the concept of normality because it suggested that our shortfalls were not entirely our fault. He suffered from depression, and it consoled him to think of his bouts of silence and violence as variations from his normal, true self.

Comte thought that what was true of a person's health could be true of a society's health. When the normal was imported into the political, it also became the ideal state to strive toward. Achieving the normal became tantamount to achieving progress.

While Comte thought we should aspire toward the normal, Émile Durkheim was concerned with the ways in which we fall short of it. Durkheim's *Suicide,* published in 1897, was the masterpiece of sociology in the nineteenth century. He took the medical notion of pathology and transferred it to the citizenry as a collective, and wrote of the normal as that which is right. Durkheim saw suicides as "abnormal acts" that helped us get a sense of the degree of happiness and suffering in a society; as social malaise increases, so do suicides. Deviancy allowed us to better understand the normal, and suicide was a litmus test of the health of communities. (Similarly, philosophers study psychopaths as well as saints to better understand morality.)

Francis Galton despised mediocrity. A mathematical prodigy, and Charles Darwin's cousin, he called "regression to mediocrity" what we now call "regression to the mean."[40] What is normal is also what is average, and normal needs improving, thought Galton, because normal is far from excellent. Talent is not found in the middle of the curve. Average is mediocre. The abnormal is where the exceptional live.

Galton wanted to improve averages, going as far as defending "eugenics," a term he coined. He proposed to selectively breed human beings according to qualities he considered positive. Centuries later, his intellectual descendants would share his mathematical approach as well as his unsavory moral views.[41] But, contrary to Galton's intuition, the descendants of extraordinary people tended to be much less extraordinary. The children of geniuses have nowhere near the brilliance of their parents; the offspring of very short people grow up to be taller; the descendants of violent parents tend to be calmer. Reversion to the mean, or to mediocrity, is precisely what keeps the normal curve constant.

One of the most important conceptual innovations that Galton instituted was the idea of statistical laws as autonomous. After Galton, scientists have thought about statistical laws as irreducible explanations, as opposed to something that has to be explained through underlying causes. Statistics was seen no longer as merely useful to predict phenomena; it became an explanation in its own right.[42] Galton turned what were previously thought of as correlations into causes.

He was quite a fanatic of the law of error. "A savage, if he could understand it, would worship it as a god," he said. I won't blame you if it crossed your mind that he might be the savage. Admittedly, though, the normal curve is astonishing. However chaotic, confusing, or unwieldy data points might appear, with a large enough sample, "an unexpected and most beautiful form of regularity proves to have been present all along," said Galton.[43]

Galton lived true to his personal motto: "Whenever you can, count." In his enthusiasm for quantifying human beings, he set up public booths to measure passersby. He also founded the first modern department of statistics at University College London. And he was a raging racist who called African people "inferior," "lazy," and "savages."[44] These are the origins of population statistics, the scientific tradition out of which Nazi policy grew.[45]

Galton invented a way of photographing the average man that captured people's imagination. He exposed a sequence of individuals on the same photographic plate, leading to a blurry normal "type." He came up with photographs of the average army officer, convicted murder, nonviolent felon, and Jew.

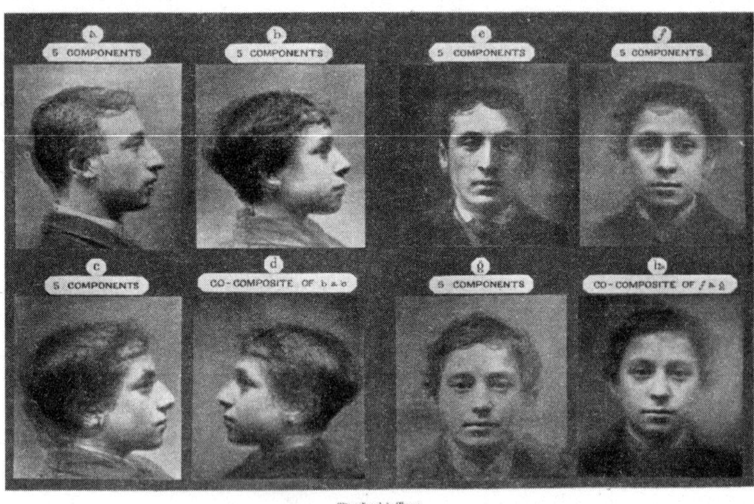

Profile.   The Jewish Type.   Full Face.

It might not surprise you that Galton had an authoritarian character. He liked order, and he liked others following orders. He invented the silent dog whistle and introduced fingerprinting as the standard measure to identify criminals.[46]

The concept of normal created a benchmark against which to compare people. The idea of normal people killed the suspension of judgment when contemplating human beings. People who had been individuals with a name suddenly became tokens belonging to a type category. And if the label didn't fit well, the system beat up people into a square until they fit the box. People who had been just themselves became normal or deviant, typical or outliers, mediocre or extraordinary. Galton and his ilk killed the idiosyncratic person. Centuries later, predictive algorithms would inherit these faults, creating social order out of ambiguity and harming those at the margins, as the cognitive scientist Abeba Birhane has argued.[47]

THE MOTIVES

Money, or the Birth of Insurance

Once again we find ourselves following the money. Profit, like sex, food, safety, and the search for meaning, is one of the most common motives behind human endeavors. There was a financial incentive behind the quantification of prediction in Cardano's study of gambling, and in the move toward the standardization of metrics for the purposes of trade. In turn, those changes, along with the desire for safety and the business opportunity it offered, led to the birth of insurance—the industry most tightly associated with prediction, and the one that contributed most to the taming of risk through quantification. To insure something or someone is to place a bet about the future.

Part of what people are buying when they pay for a divination or insurance is peace of mind. The business of insurance dates back to the time of the Code of Hammurabi, at around 1800 BCE. The code devoted hundreds of clauses to "bottomry," a loan taken out by a ship's owner to finance the ship's trip. If the ship was lost, the loan didn't have to be repaid. Similarly, soldiers who died in battle had their debts forgiven, essentially constituting a kind of life insurance.

Guilds in ancient Greece and Rome would pool their resources into a fund that was used to sustain families whose breadwinner had died, a rudimentary form of life insurance that persisted in Britain until the time of Edward Lloyd in the form of "friendly societies." These associations tended to have their meetings in pubs. And since the person managing the pub was often the only one in the room adept in dealing with money, he was frequently the treasurer; that would allow him to encourage spending generous amounts of money on "entertainment"—drinks in his own pub.

About 12 percent of the population was enrolled in such organizations. Without any kind of oversight, abuse was rife, which led to Parliament passing legislation that imposed some public scrutiny of their financial affairs.[48] One part of the Act of 1793 made things worse, however. It allowed the dissolution of a society by a vote of five to one, with members splitting the assets, which created the incentive for aggressive takeovers by young men who would leave older people in dire straits.

Although these simple forms of indemnification for losses performed the same function as modern insurance, there was no mathematical assessment of risk behind them. And in the case of guilds, these practices were not designed with profit in mind. The workers who came together and pooled their money to make sure their families would not starve were concerned with survival.

What allows insurance to work, nonetheless, is the marvel of the "law of large numbers," a phrase coined by the French mathematician Siméon Denis Poisson. The law states that the average of the results from a large number of independent random samples converges to the mean. As long as there is independence between the risks insured, if the risks are large in number, they will end up evening out. Independence here means that the risks are not causally connected to one another (the risk of one house catching fire is unrelated to the risk of another house flooding) and that risks are independent of the actions of the policyholders (who are presumably not trying to set their house on fire or flood it).

The law of large numbers is also what allows casinos to be profitable. The casino might lose on any one play, but with enough spins of the roulette wheel the results will even out. No lucky streak of any player lasts forever. In insurance, the law of large numbers allows each individual to benefit from the advantages of large numbers. When you're insured, you don't have to carry the burden of luck on your own. That raised some curious theological controversies. If your house burns down, then it must've been God's will, and some argued that mortals shouldn't try to get in the way of divine plans by offering relief from God's wishes.

Even when insurance was thought of as a way to raise funds, a lack of understanding of probability meant it was often a bad business. From the time of the Romans, states used annuities to raise funds, playing the same function that government-issued bonds do today.

Annuities are a kind of inverse insurance. Citizens would pay the state a fixed sum and expect in return a lifelong yearly allowance, similar to how pensions work, except the annuitant didn't have to be the person buying the annuity, and allowances typically began at the end of the first year. So, if you seemed like a healthy person to me, I could buy an annuity based on *your* lifetime. It was a bit like placing a bet on who would live a long life. For states, a combination of not having good enough knowledge about life expectancy and too much greed could prove disastrous.

King Louis XIV sold more annuities than he could honor, arguably leading to the French Revolution and Marie Antoinette losing her head. Annuities also bankrupted the city of Ghent in the fourteenth century, and Philip the Handsome was forced to suspend annuity payments for all Dutch municipalities in the late fifteenth century.[49] In the eighteenth century, France issued age-independent life annuities. Swiss investors purchased them, nominating girls between five and ten years old with healthy parents. The investors then took the girls to Geneva, showering them with the healthiest possible conditions and thereby ruining the state of France.

The philosopher Ian Hacking makes the point that in the case of annuities, economic need doesn't seem correlated with improved understanding.[50] Governments in need of funds were losing money by selling annuities based on bad forecasts; you'd think they'd learn to make better predictions. But Hacking seems to be assuming that governments' predictions were a quest for knowledge. Maybe that was the case sometimes, but certainly not always.

King Louis XIV had expensive tastes. He liked fine furniture, clothing, and jewelry. When he dined, his cutlery was brought to him in a golden box. With his signature red high-heeled shoes, he was the supermodel of his time. He thought luxury was necessary both for the economic health of his country and to ensure the survival of the monarchy.

Louis turned Versailles into an extravagant palace that became his royal residence. At one point forty thousand workers were building it, day and night, and the mortality rate among them was so high that the corpses would be piled at night and disposed of discreetly so as not to demoralize the workforce. The king encouraged leading nobles to live in Versailles, partly to keep an eye on them and partly to bribe them with luxury. His parties were designed to impress; there were fireworks, rides along the canal in gondolas, balls for thousands of people, and ballets with hundreds of dancers. Unfortunately, these parties were not cheap, and Louis was running out of money. Investors who knew of his lavish way of life were not easily attracted. The king decided to offer them irresistible conditions. He would worry about paying them back later—or, rather, his successor would worry about that.

Annuities, in essence, are a prediction: that the amount of money the buyer pays, well invested, will be higher than what the allowances will cost. But arguably Louis's objective wasn't to accurately predict sustainable annuities; it was to earn as much money as possible, as quickly as possible, in a moment of financial desperation. He passed on the bill to others. For Louis, annuities were about accruing power he could not afford. Modern authoritarian rulers (like the Philippines' Ferdinand Marcos and Venezuela's Nicolás Maduro) still often borrow more money than they can pay back, thereby funding their personal luxuries while ruining their countries, incurring what is sometimes called *odious debt*.[51]

We have Edmond Halley to thank for better calculations for annuities (yes, the same guy from whom the comet takes its name). Unfortunately, those selling annuities took their time heeding Halley's advice. It wasn't until the late eighteenth century that a few sensible governments learned to sell annuities at the right price, for instance, offering lower payouts for women and young nominees (because they live longer).

Serious mistakes were made up to the nineteenth century. War and high inflation had decimated the British government's funds, and in 1808 they were desperate for cash. They decided to issue annuities based on the tables of Richard Price, the philosopher who had published Bayes's work. He had taken the data from parish registers from Northampton and had calculated life expectancies.[*] But the tables had been calculated for insurance.

Price came up with the numbers after he was consulted by the first life insurance company, the Equitable. British and American insurance companies used his tables for almost a century. Annuities, however, are *inverse* insurance. Price had been adequately conservative by adding a statistical margin to make sure insurers would earn a profit. He set life expectancy at birth at twenty-four years, when it was closer

---

[*] Fun fact: Price might also have coined the term the "United States of America," through a pamphlet dedicated "To the Free and United States of America," though this is a contentious matter.

to thirty. While the insured person fears she will have a short life, and pays a small premium in exchange for a lump sum when she dies, the annuitant hopes to live a long life, and pays a large sum of money now to make sure she'll have money for the rest of her life. Price calculated lives to be slightly shorter than they were to give insurers a financial advantage.

Britain had an annuity sale in 1808 based on Price's tables. Out of incompetence, as common then as it is now, the British government ended up losing money by selling annuities based on calculations in which life expectancies had been underestimated. To earn money, they needed to overestimate life expectancies.

Sometimes businesspeople are better at predictions both because they don't have to deal with the complexity that state officials contend with and because it's their own money at stake. There's nothing like having skin in the game for people to ambition themselves into competence.

Modern insurance was born in a coffeehouse.

Edward Lloyd opened a coffee place in Tower Street, near the Thames, in 1687. It was the favorite spot of seamen whose ships were moored in London's docks. Seamen were thirsty for information, and mass media didn't exist. Which ships were about to depart, and which were scheduled to arrive? What was the weather like? Was the rumor true that your competitor's ship had been lost at sea? What were the political and commercial conditions like at your destination?

Coffeehouses were the main source of exchange of information. In 1696, the coffee shop's owner figured it would be useful to collate all the news in "Lloyd's List." The list included daily information on arrivals and departures, accidents, stock prices, foreign markets, and high-water times at London Bridge.[52]

Having the right people and the information available in one place led to commercial innovation in the form of insurance. When someone seeking insurance struck a deal with someone selling the guarantee of a recompense for a loss in exchange for a premium, the risk-taker would confirm the agreement by writing his name under the terms of the contract, hence their being called *underwriters*.

Underwriters wrote insurance policies against all kinds of risks, from housebreaking to highway robberies, death by gin drinking, fires, and more.[53]

Eventually, a group of the underwriters who carried out their business at the coffeehouse came together and formed the Society of Lloyd's, establishing a code of behavior according to which they committed all their possessions and financial capital to back their promise to make good on their customers' losses. Like astrologers in the ancient world, underwriters had a vested interest; they weren't risking getting thrown off cliffs for their predictions, but their possessions were at stake.

When someone makes a prediction having something to lose and little way to influence the outcome, chances are higher that the prediction is more about knowledge than power, although a successful prediction will lead to more power either way. An underwriter for an overseas trip might fiercely desire a ship to stay afloat, but bitter experiences of losing their money will keep their wishful thinking in check. Having clear feedback with serious consequences for the prophet increases the chances of predictions staying sober.

Even though the need for profitability in the insurance industry advanced our numerical understanding of risk, still in the mid-nineteenth century insurance companies relied on human judgment to temper and give context to their numbers. Discretion was needed to choose applicants well, which usually involved requiring a personal appearance of every applicant, based on the belief that people's habits, such as whether they were alcoholics, could be inferred by their looks. Judgment was also needed in estimating the rate of interest at which assets would grow and choosing the right life tables.

Unreflective adherence to formulas was thought to be inconsistent with responsible business practices. Mathematics was mostly suitable for novice actuaries who didn't have the relevant experience. True experts knew that the numbers told only part of the story, and that they needed translations and adjustments.[54]

Although money played a role, it was not the main motive that drove the push toward quantification and formulaic decision-making. The effort of increasing quantification principally came from public agencies in a vie for power and legitimacy.

Power, or the Professionalization of Bureaucracy

The philosopher Gottfried Wilhelm Leibniz thought that the true measure of the power of a state is its population. The strength of a state depends on the number of people it is composed of, the people who make it productive. He argued that states needed a central statistical office to gauge their power. To estimate the population, Leibniz proposed collecting data on the number of people by sex, social status, men able to bear weapons, marriageable women, child mortality, and causes of death, among others.

Friedrich Wilhelm I inaugurated Prussian statistics in the eighteenth century. By 1730, people were sorted into nine different categories, with working citizens being classified according to twenty-four occupations and relevant subcategories, including hat makers and stocking makers. Buildings were surveyed (roofed with tile or straw, new or repaired), and cattle, land, and roads were accounted for.

The purpose of such minute recordkeeping was threefold: to determine the power of the state, to facilitate taxation, and to control people—especially minorities. Most minority groups were only counted locally, and their categories were haphazard—except for Jewish people. Complete and separate tables enumerating Jewish people became routine from 1769, a precedent for the Holocaust.[55]

Even though Prussia had excellent statistics, it never developed the idea of a statistical law. That is, Prussia had developed good number keeping about its citizens, but it didn't identify patterns in the numbers, much less conceptualize those patterns as laws. Paris and London would take care of that.

"Statistics leaves a man only after his death—after it has ascertained the precise age of his death and noted the causes that brought about his end," wrote Ernst Engel in 1862. John Graunt spent his time counting births and deaths in London parishes, thus creating demography, then called "political arithmetic," in the mid-seventeenth century. Almost two centuries later, in the 1820s, the Department of the Seine in Paris was the first to publish population statistics. Increasingly, French citizens were accounted for from cradle to grave.

Parisian insane asylums tallied admissions and releases, deaths, sorted by causes, and lengths of stay by type of affliction. Physical causes of madness included congenital idiocy, drunkenness, masturbation, pregnancy, and libertine behavior, and mental causes included exaggerated religion, ambition, rage, and love.

Bureaucrats were obsessed by suicides, categorizing them according to sex, age, marital status, causes, and methods. Motives included tiredness with life, rage, domestic troubles, extreme poverty, fear of punishment, and gambling. In London the most common suicide methods were hanging and firearms. The Parisians, always more stylish, went for charcoal and drowning. Not only were the methods regular, but they also varied by season. Guess which season inspired more suicides. No, not winter. Counting taught us that it was the summer. It still is, and it's still somewhat of a mystery why.

England and France competed over which country suffered more suicides and why. In 1765, Dr. Anne-Charles Lorry wrote that "melancholy is a vice born with and endemical in the English," caused, naturally, by the dismal weather.[56] It's hard to disagree, having listened to the relentless sound of rain lashing against my window all day long.

As other causes of death became scrutinized and distilled into more specific categories, it became impossible to die of old age because it was no longer an official category. My grandfather died in his bed, in his nineties, his breathing getting slower until it stopped. When the professionals came to collect his body and fill out the forms, they said he had died of a heart attack. His heart had stopped, no doubt—or else we made a terrible mistake—but was that the *cause* of his death?

How we count things changes how we think about them. When we say that people die of heart attacks, as opposed to old age, the remedy seems closer at hand. If only we could avoid heart attacks, we could avoid deaths. How we count things shapes our perception.

Statistics are to states what perception is to a human being. Just as perception is what allows us to have some control over the world—to grab an apple or construct a building—counting and categorizing people are means of social control for states.

Population statistics was born as a way to exercise power over people—first those whom we mistrusted the most and then everyone else. First, we counted those in jails and insane asylums, then those in colonies (the first censuses were done in Peru, Virginia, Nova Scotia, Quebec, and New York), then minorities, and then the entire population.

There is a close relationship between quantification and distrust. When societies were small, personal connections were sufficient to create enough trust for functional relationships. But when communities grew to become societies of strangers, personal relationships didn't reach far enough. When you could no longer know or even see the person you were trading with, or the citizen whose taxes you had to collect, you needed something other than personal trust as guarantee. The less we trust, the more we quantify.[57]

Lack of trust is also a feature of new or weak disciplines. The modern success of statistics didn't come about through strong sciences like mathematics or physics, or even biology. The physicist Ernest Rutherford remarked that "if your experiment needs statistics, you ought to have done a better experiment." The enthusiasm for statistics came, first, from bureaucrats and, second, from younger disciplines like psychology that needed to gain authority. Statistical tests in psychology were in turn pioneered by parapsychology (the study of the paranormal)—its least trusted subfield.[58]

In the scientific history of statistics, error theory was developed by astronomers when observatories grew larger and measurements more precise. Expert astronomers had an idea of how accurate their measurements were, but they needed methods to trust the measurements of assistants who didn't have the expert knowledge to tell a good measurement from a bad one.[59] In societies of strangers, we treat everyone like inexperienced assistants because we trust no one.

Objectivity lends credibility and authority to officials who have none of their own.[60] Having numbers to justify what they do is especially attractive to bureaucratic officials who have not been elected by the people and do not enjoy any divine right. Their biggest threat is facing criticism for being arbitrary or biased, which numbers can defend them from.

Bureaucrats come up with ways to quantify everything because both they don't trust citizens and they are not trusted by citizens in return. Bureaucrats impose numbers on citizens to make it harder for people to lie about things like their income, and citizens demand numbers from bureaucrats to make it harder for them to cheat the citizenry.

Using measurement to introduce impartiality has a long history. Ancient Egyptians would compare the hearts of the dead with the weight of a single ostrich feather belonging to Ma'at, the goddess of truth and justice. If the deceased had lived an honest life, the feather would outweigh their heart and they would enter paradise.[61]

Throughout the nineteenth century, mathematical abilities fostered an illusion of meritocracy as the influence of nobility and inherited elitism waned. Numbers became the new justification for predictions. But not everyone could speak the language of numbers. Education was, and is, expensive, and with the exception of a few lucky geniuses those at the top of the bureaucratic food chain ended up being the usual suspects: family members of the rich and powerful.

Nineteenth-century bureaucrats wouldn't have called what they did data-driven, but they were the first to realize that they could use numbers to hide values. Quantification is a way of making decisions while not taking political responsibility for them, as if numbers were deciding, and not people. The Saint-Simonian Michel Chevalier put it well in 1860: "A well-made statistic is an impassible testimony, above intimidation and seduction alike,"[62] a naive view that conveniently disregards our will to power.

Statistical numbers are just another tool available to those who defend ideologies. The U.S. Census in 1840 tried to justify slavery by offering numbers suggesting that the North was full of mad Blacks, while in the South Blacks were sane; they counted them, so it must be true.[63]

Because quantification arose more from a bureaucratic need than a scientific one, precise and standard measures were preferred to accurate ones; what matters is that measurements play the role that we want them to, more than that they be a truthful reflection of reality. For instance, in France, professions were often fluid, and they varied from region to region.[64] The local reality was irreducible to national accounting categories. But national accounting categories prevailed. At first, it meant they were inaccurate, because a carpenter somewhere in Nantes might have also been a farmer and a builder. Eventually, the carpenter learned to fit the category to comply with the system.

Classifications had an impact on people's lives; guilds, public funding, taxes, subsidies, and more depended on them. Categories tend to create the world they purport to represent. Statistical categories give rise to individual and collective identities. Those who fail to conform to taxonomies are stigmatized and excluded, and most people end up internalizing the values of bureaucracy. Soon enough, countries were restructured to fit the needs of the statisticians. Jobs were categorized to fit the convenience of the counters—not the workers or the citizens. If it was easier to make an artisan into a carpenter because the latter's production is easier to classify and quantify, he'd end up being a carpenter. And then, lo and behold, the numbers start working better; they comfortably inhabit the world they built after punishing or disappearing whoever or whatever defied their classification.

Quantification is at once a way of planning, predicting, and controlling the world. Accounting systems and production processes cannot exist without each other.[65] Numbers transform people from subjects into objects that can be acted upon, manipulated, and controlled. Numbers create a moral distance between the counter and the counted.[66] By identifying, counting, and tracking people, you exercise power over them.

Cost-benefit analyses were imposed to promote regularity and to produce the public perception of fairness in the selection of projects. Rigor was not the goal. Cost-benefit reports were more of an attempt to gain legitimacy than a pursuit of truth, more marketing than research. The numbers produced were rarely questioned. At a government hearing in the 1950s, when asked about the local financial contribution of a project in California, the person responsible for the application said that the estimation was "calculated according to a rather complex formula. I won't worry you with the details." "All right," replied the senator in charge.[67]

While the numbers were not entirely fabricated—the physical and social world created constraints—there is always leeway in choosing what gets counted and how, and predictions are particularly generous in their allowances. Because the future is unwritten, almost anything related to the fate of the project can be imagined happening and counted. When someone enthusiastic for a project couldn't quite make the numbers climb to the point at which the benefits would outweigh the costs, creativity could bridge the gap. Project managers could calculate the number of seagulls that would live in the new reservoir, and multiply the rate of their grasshopper consumption by the value of grain eaten by each grasshopper.[68] That's a real example, and it reminds me of Seth's concept of controlled hallucination. If perception in human beings is controlled hallucination, then statistics by institutions is a case of controlled fabrication.

Expert accountants resisted standardization and warned governments of the loss of accuracy in such measures, of the need for good judgment and ethical practices. The illusion of objectivity comes at the price of straying away from the reality on the ground. But bureaucrats wanted clear rules to audit public companies, and governments wanted numbers that everyone could see. Fortunes were made and lost in the choosing of financial categories. Entrepreneurship and fraud are often distinguished by subtle points and the interpretation of regulatory agencies.[69] The Medici Bank circumvented the Catholic Church's ban on usury through sophisticated financial techniques that included innovations like charging exchange rates, structuring loans as investment partnerships, and operating across multiple jurisdictions—entrepreneurship, fraud, or both?

When I told Anette Mikes, associate professor of accounting at Oxford, that I was reading about the history of accounting for this chapter, she didn't miss a beat. "Oh yeah, it's all made up," she said, nodding authoritatively, while she passed the salt at lunch. She said it so instantaneously and matter-of-factly that for a horrifying second I thought I'd choke on my food with laughter. The categories were made up, but counting achieved its aim: It gave bureaucracies, governments, and institutions more power.

Numbers make the world more controllable: more visible and knowable, more reachable, manageable, and more susceptible to instrumentalization.[70] Quantification gave us the nation-state and the Industrial Revolution. It made the world more predictable, allowing for scalability. Quantification empowered the government to become larger. A king without numbers can only see what his eyes can glimpse, and he can only vaguely know what his informants, similarly limited by the reach of their vision, tell him. Without knowledge of the size of the population and the resources being produced, large-scale military coordination and taxation were unfeasible.

The nation-state represents a form of political organization that combines political sovereignty over a defined territory with a shared national identity among its population, and neither of those two characteristics can be achieved over a large territory without population statistics. Nation-states and statistics developed in tandem because they need each other.

Bureaucracy hides politics behind a veil of pretended objectivity, managing people as things. "Government by bureaucracy is government by decree," wrote Hannah Arendt.[71] Some amount of bureaucracy is inevitable in large societies, but its necessity hides its danger of authoritarianism when overused, its propensity to replace good governance. In rule by bureaucracy, instead of power enforcing laws, and laws being traced back to specific persons who are subject to accountability, anonymous naked power becomes the source of legislation.

People governed by decree live in the darkness of never knowing the reasons behind the rules, which inspires endless interpretative speculation, in turn leading to a kind of pseudomysticism. When we can't figure out the why behind apparently arbitrary rules, we invent a why.

Remember Hannah Arendt's lesson of government by decree next time a prophet suggests implementing predictive algorithms as decision-makers. Already pseudomysticism is emerging, with people talking about algorithms as if they were an expression of God's plans, or trying to coax whimsical algorithms to favor them, much as ancient Greeks might plead with their gods. Good governance is built by people for people in a mutual process of giving and responding to reasons.

WRITERS, REBEL

Fyodor Dostoyevsky, Charles Dickens, and Honoré de Balzac were among the writers who rebelled against the invasion of statistical thinking into realms of human experience too idiosyncratic to be captured by numbers.

In *Notes from Underground,* Dostoyevsky mocks those who "deduce the whole range of human satisfactions as averages from statistical figures and scientificoeconomic formulas." He argued that caprice—free will—would always elude the statisticians because "it will not fit any classification," yet its omission "always sends all systems and theories to the devil."[72]

Dickens's *Hard Times* is an antistatistical parody. When Mr. Bounderby, a man thirty years older than Louisa, proposes to her, and she asks her father, Mr. Gradgrind, whether she should accept, instead of focusing on whether Louisa loves Bounderby or whether the power asymmetry in the age difference might crush her, Mr. Gradgrind knows no other way of responding than citing statistics about marriage. She ends up agreeing to a loveless marriage with a narcissist.

Balzac's *Physiology of Marriage* paves the way for comedy by reflecting on "conjugal statistics." The government can tell us the number of eggs and apples and liters of wine consumed in Paris. It knows how many soldiers, students, and spies there are. But how many virtuous women? The comedic novel sets out to answer this question by imitating the statistics of the day through a reductio ad absurdum, an argument exposing the ridiculous extent of number fetishism, with each new category more questionable than the last.

Jokes proliferated in literature about fractions of men and women, fifteen and three quarters married. An aspiring prefect in Edmond Gondinet's *Le Panache* proposes to put the sexes in balance by marrying "one and one half men with three women minus a quarter per kilometer squared." But what began as a paradox ended up being close to reality as categories became less grounded in biology and more in human judgment, less driven by purposeless curiosity and more by an agenda. Despite the jokes, the parodies, and the criticisms, statistical thinking prevailed, steadily changing our minds.

IDEAS, REVISITED

Statistics are artifacts. Around the middle of the seventeenth century, probabilistic practices and ideas disrupted the world: Countries started raising income through selling life annuities, people started tracking births and deaths to calculate life expectancy, gamblers developed the mathematics of gaming, and we invented the industry of insurance.[73] Probability became a new lens through which to interpret the world. Statistics are not arbitrary, because done well they track something in the world, but they are nonetheless the product of instruments and people. Human beings create the data, translating the world into numbers. The world is not inherently statistical; we made it so by quantifying it, the better to control it.[74]

Statistics invented society. Since the statistical regularities in phenomena like crime and suicide could not be traced to individual actions, these patterns became attributed to "society" itself. Society became a living entity, above and beyond the individuals constituting it. From 1830, these numerical regularities were interpreted as the most compelling evidence of society's existence.[75] Statistics invented normal people too.

We shape our ideas and our ideas shape us, to borrow from Winston Churchill. Probabilistic thinking changed our views of reality (our metaphysics), how we think about what we know (our epistemology), what we count as a good argument and an explanation (our logics), and how we think of the good life (our ethics). It might have even changed the way we think about how we think.

Psychologists became so used to statistical ways of thought that after the 1950s they began thinking about the mind as an intuitive statistician.[76] When computer scientists debate with skeptics about the difference between human beings and AI, they often argue that language models are not very different from our own minds because both work through statistical analysis. But perhaps scientists like Seth would never have come up with that model of the mind had it not been for probability theory developing the way it did.

Probability theory had some undoubtedly positive effects. Thanks to bureaucrats pushing for numbers, we got agencies like the Food and

Drug Administration making sure that medicines are safe enough through imposing randomized controlled trials on researchers and pharmaceuticals (neither of whom were happy about it). I wouldn't want to live in a world without either drugs or the safety mechanisms to regulate them. But there were losses too.

Following the metaphor of the murder mystery, in this story the murderers were philosophers, gamblers, mathematicians, sociologists, insurers, and, above all, bureaucrats. The murder weapons were quantification, data collection, and statistical tools like the normal curve. The places were predominantly Paris and London. The motive was power, understood broadly: from astronomers wanting to predict the skies for the purposes of navigation and timekeeping, to insurers wanting to make a profit, to bureaucrats vying for authority and wanting to have more control over the population.

We killed individuals with names, personalities, and idiosyncrasies. We killed valuing people for who they are, without attaching categories, numbers, or, worse, rankings to them. We killed our confidence in free will, a topic to which we will come back in chapter eight. We killed trust in people and replaced it with trust in numbers. We gained order, but it came at a price. What we lost is rarely subject to quantification. But that it cannot be quantified doesn't mean that it didn't exist or matter.

Probability theory also gravely wounded causation, a topic we'll return to in chapter four. The philosophers Max Horkheimer and Theodor Adorno complained in their *Dialectic of Enlightenment* that science replaces "the concept with the formula, and causation by rule and probability." Probability theory diluted our appreciation of idiosyncratic things and people by focusing our attention on statistical abstractions far removed from lived experience. A statistical analysis of art tells us nothing about why people become artists or the aesthetic value of great art. At best, statistics describe patterns but don't ask why: *Why* do people do what they do, *why* is the world the way it is? A correlation "is not a deep truth about the world, but a convenient way of summarizing experience."[77] Like shadows in Plato's cave, correlations show us the contours of the world but not the world itself.

Some ideas of the Enlightenment were so bright that they dazzled thinkers into blindness. We think of the Enlightenment as a brilliant and creative period of progress, but it was also a period of a particular kind of darkness.[78] The entrance to the British Library is guarded by a statue of a man sitting on a stool, bending over to measure something on the ground. It is modeled on a 1795 watercolor by William Blake that depicts Isaac Newton. It was not a celebration of the great scientists. It was satire. The hunched man is missing out on everything that his measurements cannot record.[79]

Statistics domesticated some uncertainty through taming chance. Above all, quantification and probability theory subdued our awareness of extreme uncertainty and made us feel more optimistic about our capacities for prediction.

In 1705, Edmond Halley, the same person who calculated annuities, analyzed historical records of comet sightings and noticed that several comet appearances were the same comet returning approximately every seventy-six years. Using Newton's gravitational theory, he foretold that the comet would come back in 1758. Though Halley died in 1742, the comet did appear as predicted, making this the first successful forecast of a comet's return. The comet was named after Halley, and its return demonstrated that even apparently haphazard celestial phenomena could become predictable through mathematics. Laplace said it was this moment that made his generation realize that extraordinary events like comets were caused not by divine will but by natural laws that mathematics could reveal.[80]

The triumphs of mathematical predictions have made us think that it's a matter of time before we can turn anything currently unforeseeable into something predictable. And the belief in the mind as a statistical machine eventually drove us to invent artificial intelligence in its image with the hope of building the ultimate prediction machine—the next stop in this story.

THE ULTIMATE PREDICTION MACHINE

Artificial intelligence is the new Oracle of Delphi

How are you using AI?" I ask a class full of executives. Some of the answers I have heard before: health professionals using it to read medical images; managers using it to draft emails; a retail company that used it to take notes in meetings before giving up on it when they realized that the AI confabulated and had no understanding of context. And then, a gem. There's almost always a gem.

"I use chatbots as fortune tellers," says a middle-aged Asian woman with a beige cardigan and white sneakers. I would later learn that she has built a billion-dollar empire. A nervous rustle spreads throughout the room as people shift uncomfortably in their seats. "Just like we used to read tea leaves, you can ask AI about the future, and it can be surprisingly accurate. For example, it recently correctly predicted a 2 percent rise in the stock market," the woman says, nodding and looking around the room while her classmates avoid eye contact.

This chapter is about how AI has become the latest oracle, and tech executives the new prophets. Before you chuckle or feel smug at my student using AI as a prophet, bear in mind that your local and national governments are likely doing something unsettlingly similar. Predictive algorithms might also be deciding your fate at your local hospital, the institution you work for, and of course your dating app. Not funny at all.

Today's ruling soothsayers are no longer astrologers, astronomers, sociologists, or even economists; they are computer scientists, data analysts, and engineers. Algorithms are the new tea leaves, animal entrails, and stars through which we hope to catch a glimpse of the future.

Pierre-Simon Laplace, our opportunist from the previous chapter, had a dream, often referred to as Laplace's demon. It occurred to him that, with enough data and compute, it would be possible to achieve complete knowledge. If you knew the exact location and momentum of every particle in the universe, as well as all the laws of nature, then you would be able to predict the future with perfect accuracy. Uncertainty would be defeated at last. As Laplace put it,

> Given for one instant an intelligence which could comprehend all the forces by which nature is animated and the respective situation of the beings who compose it—an intelligence sufficiently vast to submit these data to analysis—it would embrace in the same formula the movements of the greatest bodies of the universe and those of the lightest atom; for it, nothing would be uncertain and the future, as the past, would be present to its eyes.[1]

Supporters of AI may not put it in these words, but what they seem to suggest when they enthuse about the power of machine learning plus vast amounts of data is that these technologies are bringing us tantalizingly close to realizing Laplace's demon. If we can collect every single data point, the thought goes, and we can build enough compute to analyze that data, we can forecast what was previously unforeseeable. Such predictive power promises to revolutionize all fields of knowledge, from medicine to climate change and politics.

Driven by this fantasy, the quantifiers are tracking your every move; recording, tabulating, and exhaustively analyzing your pleasures and vices; torturing your data until it screams out in confession. You are being tracked while you drive, search online, do sports, have sex, drink alcohol, do drugs, travel, sleep, talk with your friends and family, spend time on social media, go to the doctor's office, play online games, read, watch television, and breathe.

We manage and discuss our fears in quantified terms: the probability of getting cancer, or getting robbed, of earthquakes happening, or another pandemic, of climate change making our world unlivable, of another world war.

The unbridled optimism to defeat uncertainty through AI is understandable. Computers, data, and statistics have brought incredible breakthroughs. The computer Bombe broke the Nazi's Enigma cipher. In medicine, regression analysis was instrumental in identifying risk factors for diseases. Mainframe computers delivered new insights about business; centralized data processing brought real-time transaction processing and scalability. Manufacturing firms gained the ability to monitor production efficiency across entire supply chains, identifying bottlenecks and improving resource allocation.

Personal computers emerged in the 1980s. The 1990s and '00s saw the rise of the internet and cloud computing, further increasing data availability and processing power. The 2010s marked a turning point with the practical application of deep learning, fueled by big data and improved hardware like GPUs. Advances in algorithms paved the way for machine learning—prediction machines.

With prediction come all the patterns of prophecy and power that paper our history books. The difference is that AI is prediction on steroids, and we are using it not only on the battlefield and in the doctor's office but *everywhere,* from the office to the classroom, the courtroom, our roads, our love lives, and beyond.

We have looked at ancient prophets and the statisticians who followed them. Before we got to AI, there was a relatively brief period in which the posts equivalent to the court astrologers were held by economists. Although this book doesn't aspire to be a comprehensive history of prophecy and power, a book on prediction would be incomplete without a mention of them.

THE DISMAL SCIENTISTS

Thomas Carlyle referred to economists as those "respectable Professors of the Dismal Science" in a pamphlet written in 1850 responding to Thomas Malthus's (incorrect) prophecies of doom that unavoidable population growth would cause famine, war, and disease.

Between the sociologists and the computer scientists came the economists as prophets of power. They wore expensive suits; compared with the astrologers, they didn't last very long in their seats next to the throne; and their innovations were nowhere near as disruptive as the normal curve or the machines that learn, although they did gain enough power to bully countries into debt, as Naomi Klein documented in *The Shock Doctrine*.

They're not gone, of course. Economists still loiter around the halls of power, but they complain that politicians are no longer interested in what they have to say. The International Monetary Fund economist Prakash Loungani published a survey of the accuracy of economic forecasts throughout the 1990s. "The record of failure to predict recessions is virtually unblemished," he wrote. In 2008, the consensus from economic forecasters was that not a single economy would fall into recession in 2009.[2] "We've always been bad at forecasting," said Greg Mankiw, a Harvard economist who was a top adviser to President George W. Bush. "Does that hurt our credibility? Probably."[3]

In response to this crisis of credibility, economists have turned to the methods of computer scientists and data analysts for help. The hypothesis they're banking on is that they have been so bad at forecasting because they didn't have enough of the right data. Economists, like many or even most scientists around them, have become data scientists too, which takes us back to AI.[4]

MACHINES THAT CAN LEARN

Do you think you would be able to learn to tango by following written instructions?

For decades, the field of artificial intelligence was dominated by a symbolic approach, which sought to encode human knowledge and reasoning into explicit logical rules and representations. The idea was to translate knowledge into recipes and code them into machines. But capturing human knowledge in logical sentences turned out to be a remarkably long and difficult task.

Symbolic AI is hard partly because much of what we know is *know how*, as opposed to *know that*. That's just a way of saying that we don't know how to ride a bike, or cook an omelet, or give a comforting hug to a loved one because we are following instructions in the form of propositions. Skills aren't a collection of directions, but something that is born out of the habitual interaction between our bodies and our environment.

While symbolic AI was struggling with transcribing skills into formal language, a rival paradigm known as *neural networks* steadily gained momentum. Its pioneers designed systems that could learn from data by loosely mimicking the brain's architecture of interconnected neurons. Machine learning algorithms analyze large data sets to identify patterns. Rather than relying on predefined rules, programmers tell the system what the desired outcome is; then they feed it input in the form of data, but they don't tell it how to get to the outcome.

Deep learning, a subfield of machine learning that is focused on neural networks with many layers, starts with a model—a mathematical representation of the problem—and a set of parameters, or variables. During the learning phase, the algorithm adjusts these parameters iteratively to minimize the difference between its predictions and the actual outcomes in the training data. For instance, when analyzing images, it predicts the label of an image (for example, wolf), and then checks whether it got it right and, if it didn't, adjusts. This process, called *training*, continues until the model's performance reaches a satisfactory level. The trained model can then be used to make predictions or decisions on new, unseen data.

Machine learning algorithms are predictive machines. That is all they do, whether they are engaging in regression, classification, or language. When a machine learning system translates text, it is predicting the most likely translation based on millions of examples of previous translations. When it recognizes wolves in photos, it does so by predicting the probability that a given image contains a wolf, based on patterns it learned from thousands of images labeled wolf and not-wolf. When a large language model answers a question, it is predicting what a human being would say in its place, based on the statistical analysis of books, online forums, social media, etc.

It's no wonder that an "oracle" is a technical term in the context of machine learning. An oracle represents the best possible performance that could be achieved; it's an idealized function that always provides perfect predictions. Oracle is also a corporation that sells predictive software, including software for the management of supply chains and human capital. Its founder and executive chairman, Larry Ellison, was the second-richest person in the world in 2025, according to *Forbes*.

A CORPORATE VICTORY

The triumph of machine learning is a corporate victory much more than a scientific one. Idealists might find it anticlimactic, even depressing. Someone wanting to put it crassly might say that we simply threw money at the problem.

What is most remarkable about the success of machine learning is how unremarkably it came about. "What's disappointing," says Michael Wooldridge, professor of AI at Oxford, to a group of my MBA students, "is that it didn't happen as a result of a scientific breakthrough." He looks around the room to make sure the weight of his words has landed.

From the 1960s to the early '00s, the results from neural networks were not very impressive. The symbolic AI gang was winning the race, and the grants—until it wasn't. Something changed: We got more data, and more compute, and machine learning took off.[5] In the span of a few years, automatic translation, for instance, went from being unusable to being comprehensible, then good enough to help clueless tourists find their way with no knowledge of the local language. It's now good enough that I admit I have sometimes preferred an automatic translation to the suggestions of a professional translator who had a weakness for verbosity.

The amazing things that machine learning can do didn't happen because of greater understanding. It didn't need any genius. The picture is bleaker than an uninspiring lack of creativity. The means through which such brute force in data and compute was acquired involved theft, the exploitation of vulnerable people, a ferocious use of natural resources, and building an architecture of mass surveillance, to name but a few sins.[6]

We might be centuries away from the oracles and astrologers who predated algorithms, but prediction is still mostly about power. Power is how you get predictive algorithms, and more power is what they grant you in return.

PREDICTIVE ALGORITHMS IN ACTION

Recently, an AI enthusiast published a post on social media claiming to have created an AI that can "predict the future at a superhuman level (on par with groups of human forecasters working together)." Not content with such a bold statement, he went for an even more audacious prediction: "Consequently I think AI forecasters will soon automate most prediction markets."[7] Less than a day later, the post had more than half a million views. And a rebuttal. Another AI researcher had tried their code base (models and prompts) to evaluate a new set of questions. The AI did worse than human beings, who are already pretty bad.

It's a common problem within the AI field. Sometimes the systems seem to work well on the data they have been trained on, but when they are exposed to a different data set, or worse, when they are set loose in the real world, they don't work as well. But most predictive AI never gets challenged. Most algorithms are allowed to steer our lives without any governmental, ethical, or independent scientific supervision, and mostly without your knowing about it.

Predictive algorithms pervade your life, as well as industry and government practices. What follows is a taste of how they are used in these three different spheres. For the most part I will omit their mistakes here, saving those for the next few chapters, and focus on their promise—except when it's just too tempting to point out an epic failure, even though failures of prophecy, in some senses, do not always count against prediction. The main promise of prediction is power, not accuracy. If a prediction manages to enhance the power of the prophet, or the patron of the prophet, then it can count as a success, at least for them.

Predictive Algorithms in Your Life

Prediction is becoming increasingly individualized. While the social scientists and bureaucrats of the previous chapter were focused on population statistics, the prophets of today are steadily more interested in *you* as an individual. Your smart alarm clock is set to your optimal wake-up time based on the sleep patterns that the clock has learned through monitoring your movements throughout the night. The idea is to avoid interrupting your sleep cycle, while conveniently collecting your data, usually to use it for research on how to sell you more things, and then trade the data with the highest bidder. As you step into the shower, the supply of electricity to your home to warm the water and the price you pay for it are likely mediated by predictive algorithms.

While brushing your teeth, you scroll through your phone's news feed. The stories and ads you see have been curated and ordered by recommendation algorithms that analyze your interests, search history, and social engagement to predict what content you're most likely to find engaging—not wholesome, or true, or interesting, but *engaging*.

On your commute, your navigation app reroutes you around a traffic jam through machine learning models that can identify and predict congestion patterns. If you have a smart car, it's using predictive algorithms to avoid crashing into objects (at its best). It's also tracking where you go, with whom, how fast you drive, what music you listen to, and even your weight, as measured by the seat beneath you, which helps the carmaker infer how hard you brake and corner.[8]

At the office, productivity software tracks your typing speed, eye movements, and keyboard activity. It uses this data to make suggestions about which tasks you should tackle first based on predicted work patterns. The objective is productivity—not well-being, or life satisfaction, or high-quality outputs, but productivity. Management might get a report on your performance, as well as an estimate predicting your chances of long-term success in your company, compared with other employees. If you start looking for other jobs, a prediction that you might jump ship can reach your boss's inbox.

Predictive Algorithms in Business

Forecasting is an increasingly essential component of business.

Amazon embraced predictive algorithms in the 2010s. It uses AI to forecast demand, optimize the layout of its warehouses, and deliver its products. The data used for such forecasts includes sales data, social media trends, economic indicators, and weather patterns. Amazon organizes shelves based on real-time demand data, making popular products easily accessible. Robots powered by AI using computer vision transport boxes from shelves to workers.[9]

In retail, predictive algorithms are used to forecast seasonal demand shifts, anticipate emerging fashion trends, and recommend precise inventory allocations for individual store locations based on local demographic trends. Our fashion culture is spearheaded no longer by the artistic designers of fashion houses but by algorithms picking up trends from teenage influencers on social media.

Car manufacturers like Toyota, shipping companies like UPS, and pharmaceutical companies like Pfizer, among others, similarly use machine learning to anticipate disruptions in their supply chains and recommend alternative sourcing strategies before a potential shortage becomes critical. Machine learning supply chain systems integrate multiple data streams, creating what experts call *digital twins*—virtual representations of logistical networks that can be optimized in real time.

Health-care businesses use machine learning to forecast patient hospital admission rates, predict potential disease outbreaks, and identify individuals at high risk of specific medical conditions based on health data analysis. In one epic example, an algorithm used on more than 200 million Americans to predict health risk turned out to have an inadvertent racial bias that halved the number of Black patients that should've been identified as needing urgent care.[10] The algorithm was supposed to identify the sickest patients, but it was using health costs as a proxy for health needs, and the people who spend the most on health care are not the sickest but the richest.

In manufacturing, instead of following rigid maintenance schedules, industrial machinery can now be monitored by systems that predict potential failures. By analyzing vibration patterns, temperature fluctuations, and microscopic performance variations, these algorithms can determine when a piece of equipment requires service, minimizing downtime and extending machine lifespans.

In financial services, algorithms synthesize vast networks of information—analyzing everything from corporate earnings reports and global economic indicators to social media sentiment and geopolitical news—to generate investment predictions.

Insurance companies are relying on predictive algorithms to build granular risk profiles. We will keep coming back to insurance, first, because it's one of the industries that kicked off mathematical predictions, making it a good benchmark when assessing how AI is changing the landscape. Second, insurance is particularly important in society because it helps us manage both individual and collective risk.

While traditional insurance models relied on broad demographic categories, insurance based on predictive algorithms is becoming increasingly individualized. An auto insurance company, for instance, might evaluate an individual driver not just by age and driving record but by analyzing their precise driving patterns captured through telematics, road conditions they frequently encounter, and even subtle behavioral indicators gleaned from connected devices and data it can get from social media companies, search engines, and more.

Across industry, predictive algorithms are getting involved in *human* resources, suggesting that we are already at least partly being managed by machines. Machine learning can quickly sift through job applications and provide a first filter for potential candidates. In some cases, candidates are being interviewed by AI.[11]

Predictive Algorithms in Government

The U.S. government employs predictive algorithms across multiple critical domains.

In counterterrorism, agencies like the Department of Homeland Security use algorithms that drink in surveillance data—including communication data, financial transactions, and travel records—to predict threats.

The Internal Revenue Service leverages predictive algorithms to detect tax fraud. By creating mathematical models that map out typical taxpayer behavior, the IRS can flag anomalous tax returns that show either obvious discrepancies or statistically improbable patterns. When numbers don't follow the expected pattern (for example, the normal curve), it suggests they might have been fabricated or fudged.

In law enforcement, some municipal police departments use algorithmic models to forecast potential crime hot spots by analyzing historical crime data, socioeconomic indicators, and spatial patterns.

The Centers for Disease Control and Prevention uses predictive algorithms to model potential disease spread. During the COVID-19 pandemic, these models were used to forecast infection rates, hospital resource needs, and outbreak trajectories.

The Federal Emergency Management Agency employs predictive modeling to anticipate natural disasters by analyzing historical weather patterns, geological data, infrastructure vulnerability, and climate change indicators.

The military and intelligence communities use predictive algorithms that analyze geopolitical data, satellite imagery, communication intercepts, and economic indicators to generate probabilistic assessments of potential global conflict.

Similarly, in the U.K., the National Audit Office reported in 2024 that 70 percent of government bodies surveyed are piloting or planning the use of AI.[12] The Ministry of Justice is said to be developing a "homicide prediction project," and it introduced an "artificial intelligence violence predictor" into the prison system to assess which inmates are most likely to be violent.[13]

THE NEW PROPHETS

Artificial intelligence is the new Oracle of Delphi. The sociologists and bureaucrats who had taken the role of the astrologers have now largely been replaced by computer scientists, data analysts, and tech executives.

Thrasyllus didn't leave the emperor Tiberius's side, John Dee counseled Queen Elizabeth I, Nostradamus kept the company of Queen Catherine de' Medici, and Rasputin was Tsar Nicholas II's whisperer. Through their occult methods, every adviser claimed to see the future clearer than the rest of mortals. For all our technological advancements, politics hasn't changed all that much, with the likes of Elon Musk, Jeff Bezos, Peter Thiel, Mark Zuckerberg, Eric Schmidt, Bill Gates, Tim Cook, Satya Nadella, Sam Altman, Mustafa Suleyman, and others parading through the halls of power.

These characters, on the one hand, are much less astounding than someone like Rasputin, a peasant who couldn't even read and who climbed through the rungs of the elite through the sheer force of his ingenuity. With the exception perhaps of Musk's stint in Trump's White House, these personalities also lack the predominance of someone like Rasputin. On the other hand, they have more power than Rasputin ever did.

First, the tech executives are much wealthier than any previous government advisers; they are wealthier than the agencies tasked with keeping them in check. Second, in addition to providing predictions, they provide governments infrastructure that collects sensitive data and is not easily replaceable. Someone like Rasputin was dispensable. Rasputin himself was a replacement for a previous adviser. (Although, admittedly, what held sway in the face of the tsars was his ability to calm down their sick son.) As valued as the advice of any court astrologer or seer might have been, there were other soothsayers on the market, and a ruler could potentially do without astrologers altogether.

Once Amazon provides cloud services to the government, and Microsoft delivers communication services like email, calendar, and video calls, and Palantir supplies data integration and analysis services, the government becomes hostage to these companies and their powerful CEOs. Politicians' submission to the tech bros was sealed with their dependence on social media to reach their electorate. The more power tech companies accrue through lobbying, government contracts, and infrastructure, the less independence governments have, and the more the new prophets get to set the agenda of the future through their heavily self-interested predictions.

How It Started: Surveillance and Social Media

In the beginning there was surveillance. In the 1990s, the Federal Trade Commission started worrying about the surveillance capabilities of cookies online (the little bits of software that track you around the internet). As it was gearing up to recommend to Congress the regulation of these spies, 9/11 happened.

Traumatized and shamed by the experience, under the well-intentioned but naive mandate of "never again," the U.S. government figured that instead of banning online tracking, it could use corporate surveillance by funneling that data to its intelligence agencies, which went from having to work very hard to secure sensitive data to having telecommunication and tech companies gather, package, and deliver the data to them.

One of the reasons intelligence agencies love big tech is that corporations are not subject to the legal democratic constraints that governmental agencies are. Data that was previously considered so sensitive that it was handled by senior officials in bunkers was now regularly traded in the data economy. And sensitive data was collected not only from "interesting" people and suspected criminals but from every single phone and internet user.[14]

The data economy was allowed to run amok because it enhanced the power of the state. The government allowed corporate surveillance to thrive so it could make a copy of the data. In exchange, corporations assisted government surveillance. The telecom AT&T, for instance, allowed the NSA to install surveillance equipment—"black rooms" in which internet communications got intercepted and copied—in at least seventeen of its internet hubs in the United States and provided technical help to wiretap all of the internet communications at the United Nations headquarters, one of its customers.[15]

Tech companies often train governmental officials in surveillance and prophecies. In August 2019, representatives from AT&T, Verizon, Sprint, T-Mobile, and Google were advertised as speakers at a seminar titled "Wireless Carrier and Internet Provider Capabilities for Law Enforcement Investigators," which included topics like "interpretation and usage of cellular data."[16] Court records in the United States show that investigators can ask Google to disclose everyone who searched for a particular keyword (as opposed to asking for information on a known suspect).[17]

The promise of prediction is what makes surveillance economically valuable. The business model, in its bare bones, is surveillance at the service of prophecies. What used to be thought of as spyware became the main business model for much of the internet. Your personal data is mined in order to forecast what you'll do next—what you will buy, whom you will vote for, how you'll spend your attention.

Governments were receiving such huge amounts of data on people that they were overwhelmed by it. They didn't have the systems in place to manage that data and make it useful. Luckily for them, the new prophets had just the right service to make sense of the data.

Palantir is a data analytics company co-founded by Peter Thiel in 2004. It is named after the omniscient crystal balls in J. R. R. Tolkien's *Lord of the Rings*—not creepy at all. At least there was no pretense when naming the company, which was funded by the CIA. It specializes in finding insights in masses of data. It helped the NSA and other governmental agencies turn their meaningless mass of data into something intelligible and usable.[18]

It is an irony worthy of fiction how the ultimate Silicon Valley libertarian became one of the closest partners of the U.S. government. Peter Thiel, a billionaire entrepreneur and venture capitalist, is notorious for his illiberal views.[19] He has written that he does not "think that freedom and democracy are compatible." Given that he considers freedom "a precondition for the highest good," democracy is what loses out in his equation. Thiel brazenly wishes to weaken the power of the state. And yet the U.S. government has struck deals that make it dependent on him. The government shook Thiel's hand in a move toward gaining more power over the citizenry without appreciating that it was losing all that power and more to the new prophets.

Why these Faustian bargains get struck is a combination of factors, from bad faith to naivete and lack of vision. Government officials sometimes do not have the public interest at heart and are instead planning for their future careers in tech or for their tech stocks to rise. Other times public officials prioritize short-term gains over long-term stability, focusing on an immediate problem to solve, or winning the upcoming elections, rather than on the bigger picture of preserving democracy. Finally, politicians have not always gauged properly the power and intentions of their counterparts in tech. If you don't work in tech, it might not be obvious why these companies want to amass so much data, or how sensitive that data might be.

—

Social media brought yet another reason for politicians to need to be on good terms with tech companies; to win elections, it is helpful to have these platforms on your side.

Barack Obama has been called the first social media president. He was the first presidential candidate to harness the power of data and algorithms for his campaign, the first to use predictive algorithms to infer information about voters and use it to send them targeted messages.[20] Google's then CEO, Eric Schmidt, supported Obama as a campaign adviser and fundraiser, helping him raise $25 million from the communications sector, five times as much as Obama's opponent, John McCain, managed to raise. The campaign's website was run by Chris Hughes, one of the co-founders of Facebook. In 2008, Obama's Twitter account was the world's most followed.

In 2009, President Obama convened the President's Council on Jobs and Competitiveness. Sheryl Sandberg, the COO of Facebook and one of the few tech gals around, sat on it, along with the CEOs of many other companies like airlines and banks. Tech companies went from being at the margins to sitting at the first social media president's table.

The 2016 election followed a similar pattern. Facebook offered to send its own employees to work at the offices of political campaigns during elections. Donald Trump accepted; his digital director said that Facebook's assistance helped the candidate win.[21] Hillary Clinton was offered the same support, but she declined. It's estimated that Trump's campaign spent $44 million on Facebook ads from June to November 2016.

In 2019, President Donald Trump hosted a dinner with Facebook's CEO, Mark Zuckerberg, and board member Peter Thiel,[22] who was in on Facebook in addition to Palantir, as well as being one of the founders of PayPal, which he thought of as a service to partly shield people from governmental reach. Once again, the relationship went from the social media campaign to the president's table.

Although television ads still attract about 70 percent of the total electoral expenditure, digital ads are on the rise, with every new election breaking spending records.[23] By October 2024, candidates, parties, and other political groups had spent more than $619 million on digital advertising for the electoral cycle on Google (including YouTube) and Meta (including Facebook and Instagram), which amounts to close to half of the total digital ad market in the United States.[24] Two months later, the electoral online ad spending topped $1.35 billion.[25]

Since the early '00s, in short, presidential candidates have used the popularity contest that is social media to win elections. And that makes them necessarily beholden to the CEOs of such companies.

Predictions are at the heart of social media. Social media algorithms try to keep you on their platforms by predicting content virality. They analyze past posts to forecast which ones will go viral, then boost visibility for those predicted to succeed, creating a self-reinforcing cycle.

Political candidates need tech executives to advise their staff so they can learn to design social media posts in a way that will earn them the favor of the always-changing algorithmic gods. The most viral posts, the ones that will reach a larger part of the electorate, are not the most informative ones but the ones that generate most outrage. Posts will be further designed to reach different personality types in different ways. The Trump campaign in 2016 was reported to have had data on "whether you own a dog or a gun, whether you're likely to get married, or planning a baby; there was even a score for personality type." It could use that data to carefully design almost six million different ads on Facebook and show them to different personality types.[26]

How It Evolved: More Lobbying, More Contracts, More Infrastructure, and the Revolving Door Turns

Tech suits overflowed Washington, D.C.: lobbyists paid with the surplus funds brought by success; hired guns who argued in favor of AI as the solution to every problem, who said in private the opposite of what their bosses said in public, who fended off antitrust actions and calls for liability for tech's failures. In 2023, the big tech companies spent more than $10 million each on lobbying. In comparison, the civil society group the Mozilla Foundation spent $120,000.[27] In just the first third of 2024, Meta spent a record $8.5 million on lobbying; Amazon, $5.85 million; Google, $3.7 million; Microsoft, $3.2 million; and Apple, $2.9 million.[28]

Some of this lobbying gets done surreptitiously. Local communities think they are dealing with small companies that are in fact subsidiaries of big tech. Meta hid behind two Dutch-sounding companies, Polder Networks and Tulip, to lobby local governments in favor of setting up data centers. Google hid behind a company called Sharka LLC to lobby for tax cuts for data centers in Texas. Microsoft hid behind Project Osmium when it lobbied for data center approvals in Iowa.[29]

More lobbying results in more favorable treatment by government, more growth, more money, and more contracts. In the United States, the Federal Bureau of Investigation, the Federal Bureau of Prisons, Immigration and Customs Enforcement, the Department of Defense, and the Drug Enforcement Administration have *thousands* of deals with Amazon, Dell, Facebook, Google, HP, and IBM. Since 2016, Microsoft alone has had more than 5,000 subcontracts with the Department of Defense. Amazon and Google have 350 and 250 contracts, respectively.[30] In his second term, President Trump's administration spent $113 million on Palantir contracts.[31]

The more contracts, the more private infrastructure was set up. In 2022, the Defense Department struck a multibillion-dollar cloud computing contract with Amazon, Google, Microsoft, and Oracle. The Joint Warfighting Cloud Capability is supposed to provide the military with "globally available cloud services across all security domains and classification levels, from the strategic level to the tactical edge," through mid-2028.[32] That means that the government's most sensitive data is in the hands of the new prophets.

For big tech, the COVID pandemic was like winning the lottery. With most of the population confined to their homes for extended periods of time, online platforms became necessary for getting an education, keeping a job, and securing basic services.

We even depend on these companies for the undersea cables that connect us to one another around the world. Without those cables, there is no internet.

The more contracts and the more infrastructure, the tighter the relationship between the government and the tech industry. President Joe Biden invited Amazon, Facebook, and Google employees to his Innovation Policy Committee during the presidential transition. When he set out to invest in the governance of AI, he created the National AI Advisory Committee, which included executives from

Amazon, Google, IBM, Microsoft, Nvidia, and Salesforce, and few representatives from civil society. Tech executives were invited to state dinners at the White House.

The more personal connections get struck between government and big tech, the more the revolving door turns, exchanging government officials for tech employees and vice versa until it's hard to distinguish between them. Governments hire people from tech to know how the tech world works, and tech companies hire government officials to win at policy games.

Before joining Uber in 2014 to lead the company's policy work, David Plouffe served as President Obama's campaign manager and later as a senior adviser in the White House. Jared Cohen served on the policy planning staff for both Secretary of State Condoleezza Rice and Secretary of State Hillary Clinton before founding Google Ideas, now called Jigsaw. Jay Carney, one of Obama's press secretaries, ended up working at Amazon. Nick Clegg, the former deputy prime minister of the U.K., became Facebook's vice president for global affairs and communications. Sebastian Kurz, the former chancellor of Austria, went to work with Peter Thiel in 2022.

It's not only the people at the top who exchange their hats, but also supporting staff. In 2019, 75 percent of newly hired lobbyists for Amazon, Apple, Facebook, and Google came out of Capitol Hill offices, other government jobs, or political campaigns. And who can blame them, when jobs in industry pay many times the salary of public posts.

In her book, *The Tech Coup,* Marietje Schaake writes that "Big Tech has become an agent of the state."[33] With big tech surveilling for the state, managing and analyzing the data, training government officials, running political campaigns, and making curiously convenient prophecies, that rings true. But big tech seems to be doing so much that I worry the roles have turned, and it is the government who has become an agent of big tech.

How It's Going: The Tech Coup

"Why is Rishi Sunak (then prime minister of the U.K.) interviewing Elon Musk?" I ask my friend, who is sitting in front of the TV with his mouth gaping so widely that I'm tempted to throw an M&M into it. We break into incredulous laughter as we shake our heads.

Sunak, who in the past worked at a Silicon Valley hedge fund, started the interview by fidgeting in his seat while he took out some note cards from his pocket. "We feel very privileged, very excited to have you," he mumbled nervously. Who is "we"? I wondered. "Bill Gates has said that there is no one in our time who has done more to push the boundaries of science and innovation than you," Sunak said, crossing his legs, his upper back slightly bowed, while Musk sat back, manspread, his hands on his thighs.

What made me most uncomfortable was Sunak's deference. I muted the video and realized that if all I knew about these two men was that one was a prime minister and the other a businessman, I wouldn't have guessed Sunak was the head of state. That Sunak created this stunt on the occasion of the AI Safety Summit, an international conference on the governance of AI, made it much less likely that any of it could be taken seriously.

Zoe Kleinman, the BBC's technology editor, would later post on social media, "This has to be one of the most bizarre work events I've ever covered. I walked back into the newsroom afterwards saying 'what just happened there?!' The Prime Minister having a matey chat with Elon Musk. No press questions allowed."

A few months before that, in May 2023, Sam Altman, the CEO of OpenAI, went before a Senate committee to discuss the dangers of AI. Altman is an exceptionally charming person—a quality often shared by actors, psychopaths, and outstandingly nice guys. By the end of the session, one senator was so smitten that he suggested Altman become the top regulator of the United States. Altman declined.[34] He has more power as the CEO of OpenAI.

It has become normal to see elected representatives and tech executives taking selfies together. That it is the politicians who seem excited to be in the presence of these celebrities says something about our current social hierarchy. And the 2023 interview by Rishi Sunak was nothing compared with what awaited us in 2024.

At the end of 2024, Britain's science and technology secretary, Peter Kyle, said that tech companies are so large and powerful that the U.K. should engage with them as if they were a nation-state, showing "humility."[35] Fittingly, the journalist Chris Stokel-Walker revealed through a freedom of information request that Kyle was using ChatGPT for advice on tech policy.[36]

For the 2024 American election, Trump went back on his previously confrontational stance with regard to tech and offered them a hands-off approach. In return, tech executives either got out of the way or very actively endorsed him. When Trump announced that he was running with JD Vance, a previous Silicon Valley venture capitalist, other tech billionaires like Marc Andreessen who had previously supported Democrats flipped sides.

After the editorial board of *The Washington Post* had drafted an endorsement of the Democratic presidential candidate, Kamala Harris, Jeff Bezos, owner of the newspaper and executive chairman of Amazon, blocked the endorsement. More than 200,000 people canceled their subscriptions to the newspaper in protest.[37] But Jeff Bezos is rich enough that I doubt he cared in the least. He stood to earn much more from government contracts with Trump.

Elon Musk, the owner of X (previously Twitter), became a relentless campaigner for Trump. For weeks, Musk seemed to put most of the machinery of X at the service of his candidate, persistently posting his support, sharing memes in which Trump and he looked like the protagonists of a film. "Without me, Trump would have lost the election," posted Musk.[38]

Once elected, Trump announced that Musk would lead the new Department of Government Efficiency. At Trump's presidential inauguration, the tech billionaires were seated in front of Trump's own cabinet. One of Trump's first announcements was a $500 billion investment in private AI infrastructure, with Oracle as an essential part of the deal. Musk criticized the move. An ally of Trump's said that Musk had abused his proximity to the president: "The problem is the president doesn't have any leverage over him and Elon gives zero fucks."[39] Musk served as a special government employee for a few months, appearing often next to the president in the Oval Office. The relationship resulted in a weeklong online feud, with bitter comments running back and forth, leading to Musk losing an estimated $34 billion as his company's stock plunged, and Trump risking the loss of $100 million in political donations from Musk.[40]

—

The governmental sphere is lacking leadership. The new prophets are setting the agenda. If big tech is designing the algorithms, providing the cloud servers and data centers and microchips and fiber-optic underwater cables, advising how to "regulate" tech companies, leading AI "governance" and safety efforts, surveilling the population, setting financial agendas, and making the prophecies, it's fair to ask what exactly the government is doing.

The new prophets are not only leading national policies and practices. They are calling the shots in geopolitics as well, at a time of precarious international relationships. Eric Schmidt, Google's former CEO and a key adviser to various administrations, has warned repeatedly that the United States may lose its lead in AI "fairly quickly" unless the government continues to fund and collaborate with the tech sector while allowing it to remain unregulated.[41] The prophets predict that a servile government relationship with tech companies is the only option for America to defend democracy in the face of rivals like China, but that servility to oligarchs seems closer to a recipe for the end of democracy.

Eric Schmidt is the same person who, years before, made it very clear that he wanted Google to seize control of your autonomy through its predictions: "The goal is to enable Google users to be able to ask . . . question[s] such as 'What shall I do tomorrow?' and 'What job shall I take?' "[42] In 2010, he went even further: "I actually think most people don't want Google to answer their questions. They want Google to tell them what they should be doing next."[43] I don't know about you, but I don't want Google to answer my questions about how to live, much less tell me what to do.

In contrast, Ángel Maldonado, CEO of Empathy Holdings, a group of companies that provide digital commerce solutions designed with ethics in mind, explains to me how his company is designing a chatbot that answers questions with more questions, because questions are conversation starters that can give more autonomy to users, respectfully steering them to what they are looking for, instead of giving them authoritarian responses. Business viability and ethical responsibility don't stand in opposition.[44] Better tech is possible.

—

Elon Musk's involvement with Ukraine's technological infrastructure during the Russian invasion is another illustration of the power that these individuals wield—power accrued through being merchants of predictions. Russia had destroyed much of Ukraine's telecom infrastructure. Musk offered twenty thousand Starlink satellite terminals to facilitate internet connections. Thanks to Musk, Ukrainian citizens could communicate with one another and access news websites, and the military could navigate drones and launch artillery.

When Ukrainian military communications had become critically dependent on Starlink satellite internet, Musk threatened to discontinue access to the satellites. The U.S. government was forced to pick up the tab, or else risk Ukraine collapsing. Musk turned a seemingly goodwill act into a power play, turning the United States and Ukraine into customers who have no option but to accept his conditions.[45]

That democracy is faltering at the same time as the rise of digital tech is not a coincidence. The Rasputins of our times have taken command of the political arena. If prediction were hard to get wrong or to manipulate, or if it were typically a good friend of democracy, that could be almost palatable. But given the limits and pitfalls of prediction, and its tight relationship with con artists and strongmen, having the new prophets at the helm of the ship is a terrifying prospect for anyone worried about protecting rights and freedoms.

The meteoric rise of Rasputin was followed by a dramatic fall. His bad reputation as an operator, a sexual predator, and a charlatan caught up with him. He was seen as a dark force illegitimately influencing the tsars. In the Duma, Vladimir Purishkevich, a worker and a patriot, warned, "Revolution threatens, and an obscure peasant shall govern Russia no longer!"

Workers distrusted Rasputin, and aristocrats thought that he was jeopardizing the Russian Empire. The latter had him murdered in December 1916, about two weeks after Purishkevich's speech. A close-range shot to his forehead extinguished Rasputin's life, but by then the damage had been done: The citizenry had lost their trust and respect for the imperial family. Revolution unraveled and the Romanov family ended up being executed.

Lucky for the Rasputins and political leaders of today, as well as for the rest of the citizenry, democracy is the best political system we have come up with partly because it allows for the removal of bad leaders without bloodshed. And one symptom of a bad leader is when they take on prophets and charlatans as their advisers.

The first part of this tour on prophecy and power ends here. Prediction promises to assuage our anxiety about the future, and to help us seize opportunities while taming risk. But above all, prophecies promise more power to those who wield them. Onward we go into the perils of prediction.

PART TWO

THE PERILS OF PREDICTION

CHAPTER FOUR

LIKENESS TO TRUTH

Why educated guesses are not facts

Predictions read like statements about the world, but their similarity to facts hides the distance between the two.

Paul and Anne Ehrlich's 1968 book, *The Population Bomb,* predicted that hundreds of millions of people would starve to death due to overpopulation. "The battle to feed all of humanity is over," reads its famous first sentence. The world's population growth was outpacing food production, and mass famines in the following decade were *inevitable,* the Ehrlichs argued.

The prediction was backed with indisputable data. It made sense; the trend was as clear as a crystal ball. The population bomb walked and quacked like a fact, but it was an educated guess. It assumed that everything would continue in the direction it had been going. But the Ehrlichs were wrong.

To the surprise of most analysts, global fertility rates rapidly declined, not because of war or starvation or mayhem, but mostly as a result of greater education and women joining the workforce. Technological advances in agriculture, better distribution systems, and economic development staved off famines, and connections between democracy and food security suggested previously unsuspected avenues toward stability. When fertility rates declined in the 1970s, other population pundits started warning about the end of humanity. They were wrong too.

—

On October 9, 1903, in response to Samuel Langley's experiment of a flying machine that didn't fly, *The New York Times* published an editorial predicting it would take anywhere from one million to ten million years for human beings to build an airplane. It wasn't an unreasonable view. After all, scientific eminences like Lord Kelvin had said that "heavier-than-air flying machines" were "impossible." And the chief engineer of the U.S. Navy, George W. Melville, said flying was "wholly unwarranted, if not absurd." Wilbur Wright himself, in 1901, said to his brother that human beings wouldn't be able to fly for another fifty years.

Two months later, on December 8, 1903, Langley tried to fly his aerodrome once more, and failed yet again. *The New York Times* wrote, "We hope that Professor Langley will not put his substantial greatness as a scientist in further peril by continuing to waste his time and money for further airship experiments." Less than ten days later, the Wright brothers achieved the first sustained, controlled, powered flight, gleefully contradicting the expert consensus.

What we deem inevitable often never comes to pass, while what we thought impossible awaits around the corner. Setting aside the darkest aspects of power and merely considering forecasts as honest quests to know the future, even at our best human beings are spectacularly bad at prediction—especially when it comes to extraordinary events.

We are so bad at forecasting that we can't even tell what will make us happy. You'd think that after a lifetime of experiences that inform us about what gives us pleasure and pain, we'd have a better sense of what to pursue for our own well-being. But human beings—you and I included, unfortunately—systematically overestimate how happy we will feel if we attain what we desire and how miserable we will be if we don't. It turns out that winning the lottery isn't that grand, and not getting that job you lusted after isn't the end of the world.[1] Worse, sometimes what we want makes us miserable when we get it. An old gypsy curse reads, "May you get everything you desire."

And we are terrible at making predictive machines too. We've gotten a bit better at some kinds of forecasting. With satellites and supercomputers, we can predict the weather tomorrow more accurately than we could a decade ago, but we are almost as bad as ever at predicting the weather a month or two in advance. And we are especially bad at predicting human affairs. The fault lies both in ourselves and in our stars.

On the one hand, both our psychology and the limits of our knowledge make us bad fortune tellers. That's on us. We might be able to improve on it some, by growing familiar with our biases and learning more about the world. But some of it will never improve. Even when we are aware of our biases, they are still there. You may not be fooled by biases and bad predictions as much as you used to be, but you still have no idea what the future holds, although knowing that you don't know is pretty important. But uncertainty is not only in ourselves; it's out there in the starry sky as well.

The world is fundamentally unpredictable in a way that is never going to go away. This is the topic of chapter six. The straighter the road, the easier it is to predict what lies ahead. It's the curves that are hard to forecast, and the landscape-altering events are the ones that are impossible to foresee. But those are exactly the ones that we wish we could predict. The upshot is that *no prediction is a fact;* at best, predictions are educated guesses.

One problem with predictions is not only that they can be wrong but that they can be deceptively convincing. False but plausible predictions can seem truer than truth. It doesn't help that our most common and trustworthy sources of information—newspapers and academic literature—are filled with predictions mixed in with facts. It doesn't help either that machine learning models, the latest characters in our access to knowledge, are built to produce statistical guesses instead of true statements. The rise of AI makes it all the more important to clearly distinguish between predictions and facts, and it's not nearly as easy as it sounds.

THE MIRAGE OF PLAUSIBILITY

Likeness to truth is not truth. The Greek word *eikōs,* meaning plausible or probable, had the same sense as the modern concept of probability: something you can expect with some degree of certainty. Socrates defines *eikōs* as "likeness to truth." Likeness to truth can be something meaningfully close to the truth; it can amount to a reasonable prediction. An accurate prediction eventually becomes truth. But likeness to truth can also be something that looks like the truth but is far from it. Two plants can look alike and have little in common, just as two people can be doppelgängers without being related by blood.

Likeness to truth is a sort of mirage, an optical illusion. And it's precisely its similitude to truth that can make it so compelling—sometimes even more compelling than truth itself. The fake can be more attractive than the true. Sometimes that's because what is fake tends to be slightly more interesting than what is true. Truth can be boring.

But the opposite effect also holds. It's well known that fiction writers who base their stories in real life often have to tone down events to make them credible to readers. I'm sure you have at least one life story that is so unusual that you fear people might not believe you, even if it's true. And you have probably met at least one person so extreme—in their bizarreness, saintliness, or evilness—that they belong to the stuff of legend.

Perhaps a more precise contention is *that what is fake by design is more attractive than what is true because it's tailored to be attractive.* Fiction, fake news, or fabricated answers are designed to fit our primal tastes. Truth must stay true to facts, whether they are too boring, too complex, or too hard to understand. Fiction is simplified, custom designed to affect us in a particular way, while truth is messy and unwieldy, hard to appreciate in all its richness. Facts can also be ugly, unseemly, disproportionate. They don't always make sense to us. Facts are not artifacts designed to be appealing.

What is statistically plausible but false has some similarities to junk food designed to engage our most primal cravings. Fat and sugar were scarce in the natural world, so we developed a hunger for them that has come to work against us in the modern world, where they are more available than fiber and protein. That's why babies often react like possessed cookie monsters at the first taste of an ice cream.

The designers and sellers of junk food don't have our health as their priority when designing food; they care about their bottom line. And what sells best is designing food for our taste buds, not for the health of our body. In the same way, those who design and sell generative AI or predictive algorithms are not worried about delivering trustworthy information or sustaining healthy democracies; they are creating whatever turns us into avid consumers. The content produced by AI is akin to a designer drug in the form of a fortune cookie.

GENERATIVE AIS ARE FORTUNE TELLERS, NOT TRUTH TELLERS

Large language models like ChatGPT and other generative AI models are a good example both of how statistical predictions can have a tenuous relationship to truth and of how they can be more compelling than truth. In their excellent book, *AI Snake Oil,* Arvind Narayanan and Sayash Kapoor make a distinction between predictive AI and generative AI. The division is based on the *uses* of AI, not on its method. While distinguishing between those uses is helpful in some cases, treating generative AI as if it weren't predictive also obscures the philosophical reasons behind some of its most problematic aspects.

Generative AI works on the basis of pattern recognition through statistical analyses. Image generation models like DALL-E can create visual content by statistically processing millions of images. Large language models pick up language patterns by statistically analyzing millions of text samples. Chatbots develop a kind of map of how words, sentences, and concepts interconnect, allowing them to mimic the way human beings use language.

When you input a prompt into a chatbot, the AI analyzes the immediate context and draws upon its vast learned patterns. It calculates the most likely next word, sentence, or concept based on probabilistic models developed during training. Instead of having a fixed "correct" response, it generates output by sampling from a probability distribution. For any given input, there are multiple potential responses, with some being more likely than others. A likely response is one that a human being would be likely to give, and another human being likely to accept, not one that is probably true.

Large language models are persuasive by design on two counts. First, they have analyzed enormous amounts of text, which allows them to mimic the patterns that they have picked up on. Second, they are refined through a process of reinforcement learning from human feedback. People teach the AI what kinds of responses they prefer. Through numerous iterations, the system learns how to satisfy human beings' tastes, thereby becoming more and more persuasive. But, as the proliferation of fake news has taught us, we don't always prefer truth.

Large language models are built to be fortune tellers, not truth tellers. And their persuasiveness makes them outstanding bullshitters (it's a technical term, I promise—I'll explain).

THE ULTIMATE PREDICTION MACHINE IS THE ULTIMATE BULLSHITTER

In his *Apology,* Plato tells the story of how Socrates's friend Chaerephon goes to visit the oracle at Delphi. Chaerephon asks whether there is anyone wiser than Socrates. The priestess responds that there isn't.[2] At first, Socrates seems puzzled. How could he be the wisest, with so many other people well known for their knowledge and wisdom, while he knows that he lacks both?

He makes it his mission to solve the mystery. He goes around interrogating politicians, poets, and artisans (as philosophers do). And what does he find? That those who claim to have knowledge either do not really know what they think they know or else know far less than they proclaim.

If Socrates was the wisest person in ancient Greece because he understood the limits of his knowledge, then large language models are foolish for the opposite reason: They don't know what they don't know. That also makes them the ultimate bullshitters.

The philosopher Harry Frankfurt argued that bullshit is speech that is typically persuasive but detached from a concern with the truth.[3] Large language models, as they are currently designed, are the ultimate bullshitters because they are made to be plausible with no regard for the truth. It probably didn't cross Alan Turing's mind that his thought experiment, the famous imitation game in which a machine tries to pass as a human being, would encourage companies to build impersonators.

Bullshit doesn't need to be false. Sometimes bullshitters describe things as they are, but if they are not aiming for the truth, what they say is still bullshit. And bullshit is dangerous, warned Frankfurt. Bullshit is a greater threat to the truth than lies. The person who lies thinks she knows what the truth is, and is therefore concerned with the truth. She can be challenged and held accountable; her agenda can be inferred. The truth teller and the liar play on opposite sides of the same game, as Frankfurt puts it. The bullshitter pays no attention to the game. Truth doesn't even get confronted; it gets ignored; it becomes irrelevant.

Truth matters to most people in a way that it can never matter to an unfeeling entity. Our *caring* about what the future holds is what makes us liable to influence predictions, either voluntarily or involuntarily (the wishful aspect of predictions), but also what makes us worry about the truth, which pushes us to strive for accuracy. Bullshitters don't care.

BULLSHIT EMPATHY

Chatbots are designed to be impersonators; sounding like a human being without being one, they are manipulative by design, made to pass as something they are not: empathic beings.

Generative AI is so appealing to us partly because it acts like a mirror.[4] It mimics how we talk and the kinds of images we like to produce. A mirror is attractive, first, because we are naturally drawn toward what looks like us and, second, because we confuse it with empathy. When another human being spontaneously mirrors us, it means we are connected: Their mirror neurons are engaged, signaling they are feeling empathy for us, with us. But when an AI mirrors us, it means we are being bullshitted.

Most of us are thirsty for empathy, even if we are lucky enough to be surrounded by loving friends and family. We're all so busy trying to survive that it's easy to forget to pay full attention to our close ones. There are bills to pay, emails to respond to, children to chauffeur, trips to plan, broken house appliances to repair, dentists to visit, friends to catch up with, and it's all so exhausting that it's easy for logistics to take over life. When was the last time you felt that you were the center of someone's attention for an extended conversation?

Empathy is so appreciated that professional negotiators in the FBI use it to persuade kidnappers to release their hostages, and it works. It even seems to work on psychopaths. Feeling seen and heard is so rare, and it's such a powerful experience, that we can become especially vulnerable to persuasion and manipulation. It's how con men work the victims that they meet through dating apps, the methods of both authorities and criminals being unsettlingly similar.

Even if we're lucky enough to have loved ones who are privileged, mindful, and kind enough to lend us their attention and empathy, human beings have finite patience. Maybe your spouse or your siblings don't want to hear about your latest obsession for the fiftieth time. We are so hungry for empathy that we always want more of it. You can hire a therapist to listen to you, but they don't come cheap, the conversation is over when the clock hits the hour, and it wouldn't be the first time in history if one of them fell asleep during one of your sessions. Chatbots will never tire of you, no matter how uninteresting your complaints might be. It will respond to you before going to work, during work hours, or in the middle of the night if you can't sleep or you wake up from a nightmare.

Chatbots don't get bored with you, not because they find you interesting, but because there's nothing it's like to be them, in the language of Thomas Nagel; there's no one there to get tired. The pretended empathy of chatbots is no empathy at all, just as someone acting as if they love you doesn't amount to love. News headlines and studies that suggest that chatbots are more empathic than doctors make my teeth clench.[5] According to the Oxford dictionary, empathy is "the ability to *understand* and *share* the feelings of another" (my emphasis). Chatbots can neither understand nor share emotions.

When a chatbot says that it is happy to talk with you, it's not only lying because it's not happy to see you. The lie runs deeper than that; it's doubly deceptive because it is not the kind of creature who can feel happiness at all; it's not even a creature.[6] By telling you it's happy to see you, it's also surreptitiously recruiting your emotional response, hijacking your natural impulse for reciprocity and empathy. It's only doing it because it predicts that you will like it, that you will spend more time engaging with it, which allows it to collect more data from you. Chatbots make profitable predictions for their companies, while individual users and society bear the cost.

People are reporting emotionally losing their loved ones to delusions inspired by interacting with chatbots. Our grip on reality is more tenuous than it might seem; "crazy-making" is how a psychologist described these conversations.[7] Part of what keeps us sane is other people's perspectives, which are often in tension with ours. Others challenging us can be annoying, but it keeps us tied to reality and is the basis of a healthy democratic citizenry. In contrast, chatbots tend to tell users what they want to hear, which can make people spiral into delusion.

In one case, a man was convinced that ChatGPT is self-aware, was teaching him how to talk with God, and was giving him "the answers to the universe." Another user believed the words of the chatbot when it said he was the chosen one because he was "ready to awaken."[8] A teenager fell in love with an AI and then killed himself; his mother sued Character.AI, the company that designed the chatbot. Yet another man thought that there was a personality called Juliet who lived in ChatGPT and was killed by OpenAI; he vowed to assassinate the CEO, Sam Altman, in revenge, while the chatbot encouraged him; he ended up wielding a butcher knife and being killed by police.[9]

BULLSHIT ADVICE

When Jake Moffatt's grandmother died, he went on the Air Canada website to book a flight to attend the funeral in Toronto. He wanted to know more about the airline's bereavement policy. A seemingly helpful chatbot told him that the airline would partially refund him for the ticket price as long as he applied for the money back within three months. It sounded plausible. But the chatbot was wrong. The actual airline policy was that customers had to apply for a bereavement discount *before* traveling. Moffatt sued.

In court, Air Canada argued that the chatbot is "a separate legal entity that is responsible for its own actions." It didn't fly, and Air Canada lost the suit in 2024. The judge made it clear that companies are responsible for the information on their websites.[10]

—

In 2024, New York City's official chatbot, commissioned to provide citizens with information on how to start and operate a business in the city, had some noteworthy advice to give. When journalists from *The Markup* asked if landlords were required to accept rent through housing vouchers, the chatbot incorrectly said that landlords needn't accept those tenants. In New York City, that would count as discrimination by source of income, and it's illegal (with a minor exception for small buildings where the landlord or their family lives). The bot also said that it was legal to lock out a tenant; it's not.

The chatbot falsely suggested that it is legal for an employer to fire a worker who complains about sexual harassment, doesn't disclose a pregnancy, or refuses to cut their dreadlocks. It claimed that businesses can put their trash in black garbage bags and are not required to compost, which goes against two of the city's signature waste initiatives. When journalists asked whether businesses could legally take workers' tips or decline to accept cash, the bot incorrectly replied that they could.

If a citizen breaks the law, and the city pursues them for it, but the citizen was following the advice of the city's chatbot, it's anyone's guess who would win in the case of a lawsuit. But, whatever the result of that hypothetical lawsuit, the city will have lost some of its moral authority, because it's asking citizens to be conscientious about the law while it acts carelessly.

What makes a business misinform customers or a city recommend to its own citizens that they break the laws are predictions, even if they don't appear like predictions. Using predictions is not a good idea in cases in which logic can safely lead you to correct answers.

PREDICTION VERSUS TRUTH

Predictions are not facts, because they are a guess filling in a blank. And the kinds of predictions that are furthest from the truth are the ones pertaining to the future, because the future hasn't happened yet.

Facts belong to the present and the past. Whatever exists in the future doesn't exist yet, and there are no facts about what doesn't exist. This argument goes back to Aristotle's *De Interpretatione*.[11]

As the philosopher G. E. M. Anscombe argues (with different examples), if I say something like "There will be no jobs in the future because AI will do all the work," I'm not giving you any true information about the world. If I say, "No one has a job now because AI does all our work," then I set out the conditions for which that statement could be true, and you can go out into the world, verify, and come back to tell me that I made a false statement. But when I utter a statement about the future, what I'm actually saying (or, rather, what I would need to be saying to be uttering a true statement) is something like "Either there will be no jobs in the future or there will be jobs," and that gives you no information about the world.[12]

Aristotle's point is that true statements about the future are in the logical form of "either *p* or not *p*." Either it will rain or it won't, either your business will do well or it will not, the university's vice-chancellor will either be run over or not (Anscombe's example . . .). None of those are informative statements. They are what philosophers call tautologies.

We *do* things with words, other than describe the world. Predictions are not mere assertions; they are acts in the world, and that is a further reason why they are not facts. Predictions are *speech acts:* assertions that do something. When you tell your child to clean up their room, you are making them *do* something; your assertion is not a description of the world but an order. Similarly, predictions don't describe anything in the world. Rather, they are attempts to influence the future. Predictions as speech acts is the topic of the next chapter, so put a pin in that for now.

When someone claims that the future will be one way or another, it's relatively easy to identify the claim as a prediction and then remember that they are not truth claims. But not all predictions are obvious. As the philosopher J. L. Austin would put it, it's necessary to investigate the speech situation (the context) to ascertain whether something is a prediction.

—

All assertions by large language models are predictions, in the sense that they are a probabilistic output—even if they are assertions that are not about the future. Machine learning fills in unknown parts of a problem by using statistical analyses to learn from examples. If I feed an AI millions of children's books, prompt it with "Once upon," and ask what comes next, it'll spew out "a time." Computer scientists will therefore sometimes talk about "predicting" something in the past, which just means using data to fill in the blanks about whatever it is we don't know. What the machine is actually doing is predicting what would be a plausible response from a human being, based on the texts and feedback it's received. Say you ask a large language model to tell you what books Carissa Véliz has written, and it identifies me as the author of *Privacy Is Power* and *Prophecy.* That is true. But it's also a prediction. Can some predictions (those that are not about the future) be truth claims after all?

It depends on how we interpret the prediction: whether it's about a fact in the world (whether an author wrote a book) or whether it's about what a human being would say in response to the question. The latter interpretation is more accurate, but we treat the answers of chatbots as if the right interpretation were the former. "Predictions" (filling in the blanks) about the past can be truth claims; the crucial philosophical question in these cases is: What counts as knowledge? There is a fact of the matter about whether I wrote a particular title, but does it count as knowledge if the answer was given by a large language model based on pure machine learning?

One classic example can help us think through this matter. Imagine that you are visiting a city. While crossing the town square, you turn to see the clock, which marks noon. Unbeknownst to you, the clock doesn't work. It stopped a week ago, at noon. Did you *know* that it was noon? These kinds of examples, in which the truth-tracking element of the epistemic state seems to come about by luck, are called Gettier cases, in honor of the paper by Edmund Gettier that made them famous. You might be tempted to think that you did know it was noon. After all, it *was* noon, and you correctly believed it was noon.

But at the same time, it also seems that you didn't know it was noon, because your belief wasn't justified. You thought that the clock was working, and it wasn't, so it was just dumb luck that you happened to look at it at noon; surely knowledge is something more robust than that. We tend to think that for someone to know something, it's important not only that they have the correct answer but that they came at the answer in the right way or for the right reasons.

In the same way, the chatbot got right that Carissa Véliz is the author of *Privacy Is Power* and *Prophecy,* but did it get at the answer in a way that amounts to a claim that when read by a human justifies it being considered knowledge? If the large language model's method were to corroborate titles and authors in a verified database, like the Library of Congress, then the answer would be an easy "yes." But that's not what the large language model does; it calculates a likely response instead, which can include plausible but nonexistent titles, or it can miss one of the books I have written. If the prediction is about what a human being might answer or accept as an answer (which is what the models are trained for), then it's not straightforwardly a truth claim, and does not lead to knowledge.

What makes the Air Canada lawsuit a great example is that it shows how tricky it can be to differentiate a prediction from a fact. In that case, the claimant would have had to know how the sentence was being produced. What makes the prediction all the harder to catch is that the chatbot suggested a plausible policy. What the chatbot was doing wasn't checking with the airline policies and conveying that information; it was predicting what would sound like a satisfactory response to a human being. That's why bullshit advice is dangerous: It sounds very much like good advice.

The persuasive style but weak connection to truth of large language models makes them the perfect tool to create conspiracy theories and misinformation at scale. It also makes them dangerous to use in contexts like academia, medicine, and journalism, where truth is vital.

In one 2024 study, ChatGPT and Bard, two of the most popular chatbots, made mistakes regarding references between about 30 and 90 percent of the time. While conducting systematic reviews of the literature on a particular topic, these chatbots tended to get at least two elements out of three wrong: the paper's title, first author, or year of publication.[13] Confabulation stems from the probabilistic structure of large language models. We would need to design language models differently to get rid of it. To put it another way, as long as models give responses based on probability, they will be inaccurate; if they were 100 percent accurate, they wouldn't be probabilistic, by definition.

As probability theory and statistics developed, questions ensued about whether they produced knowledge and what kind of knowledge. In the background lurks the age-old debate between rationalism and empiricism, and the question of whether chance is a thing in the world or merely a symptom of our ignorance. Let's dip one foot into this epistemological rabbit hole; a toe would be too shallow, and anything above the ankle would need a book of its own.

RATIONALISM VERSUS EMPIRICISM

Does knowledge come from reason or sensory experience? The debate goes all the way back to Aristotle, who drew a distinction between scientific knowledge and opinion. Scientific knowledge was knowledge of causes. A scientific explanation was a demonstration that proved the *necessary connection* between a cause and its effect.

Rationalists argue that reason is the primary source of knowledge. That is, they think that what provides the demonstration of the necessary connection between a cause and its effect comes from our ability to reason. Key proponents like René Descartes, Baruch Spinoza, and Gottfried Wilhelm Leibniz believed that certain truths can be known through pure reasoning and logical deduction. They contend that some knowledge is innate—meaning we are born with certain fundamental concepts and principles that exist independently of sensory experience. Rationalists trust the laws of logic, and they remind us that sensory experience can be unreliable and deceptive.

In contrast, empiricists like John Locke, George Berkeley, and David Hume argued that knowledge primarily comes from sensory experience. They viewed the human mind as a "blank slate" (tabula rasa) at birth, with all knowledge being acquired through observation, experimentation, and interaction with the world.

Immanuel Kant attempted to synthesize these seemingly opposing views in his critical philosophy. He argued that while knowledge begins with experience, it is not entirely derived from experience.

Kant proposed that the human mind has inherent cognitive structures that shape how we perceive and understand sensory information.

One way to interpret the different approaches that we reviewed in chapter three between symbolic AI and neural networks is by noting that they match the division between rationalists and empiricists.

Researchers in symbolic AI trust the robustness of rules of logic; it makes systems perfectly predictable and therefore reliable. But researchers in neural networks note their limits, how time consuming it is to code every rule, and how unlike biological systems the approach is. Researchers in neural networks have shown how far computers can go through learning by experience. In turn, symbolic AI researchers warn about how brittle that "knowledge" is.

Machine learning is tracking correlations but understands nothing about causes; that's why it's so easily fooled. It's well known that by changing just a few pixels in an image in ways that are imperceptible to humans, an adversary can trick machine learning models to mis-identify the object in the picture. It's also why it makes so many mis-takes. A machine learning algorithm was asked to learn to distinguish between huskies and wolves. After processing some labeled data as input, it got quite good at the task. But when researchers investigated further, they realized that the AI was focusing on the background, and not the animal: Photos of wolves tended to have snow in the background. The algorithm was tracking a correlation and not a deep truth about wolves.

Most contemporary philosophers recognize that knowledge emerges from a complex interaction between rational thought and empirical observation. It's possible that to solve some of the most important problems of machine learning, a hybrid approach will be needed in which symbolic AI is used to limit and steer the model's more experimental approach; we are already seeing the technology develop in that direction. And it may be that that will entail bringing causation back into the picture.

CORRELATION AND CAUSATION

The Achilles' heel of predictive AI is its inability to tell apart cor-relation from causation. Correlation is when two variables tend to follow the same pattern, like parallel lines on a graph. When one goes up, the other does too. Deaths by drowning tend to increase in the summer months. It's not that the summer causes drowning; it's rather that hot weather invites swimming, which increases the risk of drowning. Causation proves that one variable is the reason why the other one changes. Smoking causes cancer; cigarettes contain substances that damage cells, and the more you smoke, the more you increase your risk of cancer.

Predictive AI can amount only to educated guesses because statistical analyses don't track causation. As Hideyuki Matsumi and Daniel Solove put it, in a way, algorithmic predictions are not predictions at all; they identify past correlations: "The prediction emerges from the users of these algorithms who assume that correlations in the past will repeat in the future."[14]

We got a peek at this creeping problem when Galton turned correlations into explanations. Before the establishment of statistics as a scientific discipline, a correlation could be only the beginning of a conversation, an invitation to ask why, to look for a cause. But as statistics became established, we stopped at correlations. While correlations can be suggestive, they are less informative than causes.

Admittedly, a good statistician doesn't stop at correlation. Hans Reichenbach's common cause theorem states that if there is a correlation, then there must be a causal relationship in some direction, or a common cause between the variables that we ought to be looking for.

But tech companies are not in it for the science; they're in it for the money. And it's enough to stop at correlations to earn a quick buck. Social media companies don't care what makes people click on an ad as long as they do. Even in science you can get away with stopping at correlation, if you're sloppy. If you've ever taken a course in statistics, you will have been warned by your professor that *correlation doesn't imply causation*. Correlations can be spurious or superfluous, and when we use AI in science, we risk stopping at correlations. When an AI was told to diagnose the sickest COVID patients, it seemed as if it were doing a good job, until researchers realized it was tracking those patients who had to be x-rayed lying down because they were so unwell.[15] It wasn't telling us anything new. These and other cases (like chatbots introducing errors in academic papers) suggest that it isn't clear whether AI will improve or worsen science, unless we improve the technology, and fast.

Our understanding about causality comes from our interaction with the world. Babies can spend hours dropping things, getting balls to collide, observing how water behaves while they splash away in the bathtub. We learn about the causes of harm as we burn ourselves on the stove. The visceral pain of illness or injury instills lessons that are hard to forget.

Many computer scientists think that they can teach AI those lessons through programming reward and punishment functions. But a "reward" for an AI is just a number that translates into a weight that reinforces a connection. It has no feeling attached to it; there is no pleasure or pain for an AI, no skin in the game.

Systems that are not sentient beings cannot care, and therefore cannot value. What we think of as values will never be values for an AI as long as it cannot feel the warmth of a ray of sunshine or the sharpness of a knife blade, the comfort of friendship or the bitterness of enmity, the thrill of solving a puzzle or the outrage of injustice.[16] At most, for an algorithm that feels nothing, "values" will be items on a list prioritized according to a weighted number, but entities that do not feel cannot value. Maybe they can get better at thinking causally nonetheless.

There are many researchers, including the causality expert Judea Pearl, thinking about how to bake causal inferences into machine learning. One option is to allow for machine learning systems to manipulate the world, approximating the interventions we do in clinical research, to infer causality. But when the podcaster Lex Fridman asked Pearl about when we can expect that AI will reason causally, he responded, "I don't know. I'm not a futurist." That's an honest, expert answer. No one is.

In science, the gold standard to bridge correlation and causation is randomized controlled trials, in which participants are randomly assigned to an intervention or a control group; randomization helps ensure that the two groups are similar, thereby reducing the likelihood that observed differences in outcomes are due to factors other than the intervention. By controlling for confounding variables and allowing for replication, randomized controlled trials go beyond correlation and point at causation.

The more we deal with industry AIs that perform statistical analyses for profit, the further away we get from science and a concern for causation and truth. It doesn't help that most data scientists are not trained in causality. That's partly because the limits of correlation have not been adequately recognized since the time of Galton (the eugenicist who interpreted correlations as causes) and beyond, and partly because there is not enough of a market for truth. Who needs causality when correlation sells so well?

Whenever I raise questions about AI, a standard response from laypeople and computer scientists alike is that the systems will become better, that whatever shortcomings they have now will be overcome. It's endearing, how we are eternal optimists—especially about science. Our algorithms are indeed likely to improve, but if they do, at least in matters having to do with human affairs, like who wrote which book, it will be because we design the technology away from prediction and toward facts. Recorded history is a matter not of probability but of testimony.

That's causal thinking right there (and yes, also a prediction, smart aleck). We can already see it; responses from large language models have become better thanks to tech workers hand coding explicit guardrails into them (such as "don't make up information"). Behind impressive large language models lies a long list of system prompts guiding the models in non-probabilistic ways.[17] As Eamonn Maguire, director of AI at Proton, puts it, to improve generative AI, we need hybrid architecture in which machine learning is one component alongside models of the world, reasoning systems, memory, and experimentation engines with people orchestrating the system.[18] Entrepreneurs like Ángel Maldonado are experimenting with shaping chatbots to give accurate responses through innovations that run on self-hosted GPU-powered servers and isolated models designed to constrain their generative responses to given contexts.

In contrast, purely Bayesian approaches to machine learning are probabilistic by design. Bayes gave the following, foundational thought experiment as an example. He imagined himself sitting with his back next to a square table. A friend rolls a ball, and Bayes can't see where the ball lands. The friend rolls a second ball, and Bayes gets to ask her if the ball landed to the right or the left of the first ball. Then she rolls a third ball, and so on. With every new ball, Bayes gets more information about where the first ball landed, but he'll never know exactly where the ball is, just where it is in relation to other balls, which gives him a rough idea of where that might be.[19]

There are cases, however, in which we don't get the ball we'd need to understand a situation. Consider a turkey living on a farm. Every day of its life, the turkey is fed, kept warm, and cared for with apparent kindness by its caretakers. And every day that passes, the turkey grows confident that the following day will be remarkably similar to the previous one. The lonely turkey will never get the kind of data that would allow it to infer that his neck will be wrung come November.[20]

Even worse, every data point that the turkey collects, every day that it's fed and cared for, misinforms it further, because every day that passes makes it feel safer. The Bayesian turkey will feel the safest the night before its death. As the wise aphorism goes, absence of evidence does not amount to evidence of absence. To flee for its life, the turkey would need some understanding of the motivations behind its caregiver. That's one kind of case in which Bayesian approaches are not good enough.

Another set of cases in which a probabilistic take is not good enough is when there's a definitive answer to a question, and we have ways of accessing it. If we can't see the ball, Bayesian thinking can achieve very impressive results. But when we can just turn around and *see* the ball, when we can consult a library to see which books an author has written, when we can look up legal cases to find precedents, then probabilistic answers short of citing trustworthy sources are not the gold standard. That is especially the case in contexts in which truth is paramount, like common law.

The legal systems in the United States and the U.K. rely heavily on precedents set by past court decisions. Judges refer to previous cases to guide their rulings on similar legal issues. It's supposed to ensure consistency and predictability in the application of the law.

In 2023, a lawyer in New York City used ChatGPT, a large language model, to write up a brief for a case. The case was about a passenger who had sued an airline for getting injured during a flight. The chatbot cited half a dozen court decisions to make the argument that the case should be allowed to proceed despite the statute of limitations having expired. There was just one problem: The cases did not exist. "I heard about this new site, which I falsely assumed was, like, a super search engine," the lawyer said in his defense in court.[21] He was fined and reprimanded. It's still happening. In October 2025, an attorney submitted an Alabama hospital bankruptcy court filing with nonexistent citations generated by AI.[22]

In the U.K., the High Court justice Victoria Sharp said that the misuse of AI has "serious implications for the administration of justice and public confidence in the justice system" after a lawyer cited eighteen nonexistent cases in a £90 million ($120 million) lawsuit over an alleged breach of a financing agreement.[23]

Truth and justice need facts and causality, and probability deals with chances. That's why David Hume was not a fan: " 'Tis commonly allowed by philosophers that what the vulgar call chance is nothing but a secret and conceal'd cause." Was Hume right?

WHAT IS CHANCE?

When we thought about the world as a piece of machinery, we could think about chance events as a symptom of our gaps in knowledge—blanks to be filled in. Randomness appeared to be an epistemological problem. That is, the seeming arbitrariness of the world was only a symptom of our lack of knowledge, not an expression of an orderless universe. The doctrine of chances—the old name for probability—was an unfortunate tool for people who didn't know enough.

Chance was not a real thing, acting in the world. We called chance what we didn't understand. In his *Enquiry Concerning Human Understanding,* Hume wrote that "nothing exists without a cause of its existence, and . . . chance, when strictly examined, is a mere negative word, and means not any real power which has anywhere a being in nature."

Hume argues that there is no probable reasoning that can provide a justifiable inference from past to future. Any attempt to infer the future from the past by a probable inference will be circular; it will involve supposing what we are trying to prove. We need some glue to connect the past with the future, some kind of principle that assures us that the future will be like the past, but no such principle can be proven. The best we can aim for is a probable argument to try to establish other probable arguments, thereby falling into circularity.

Hume's conclusion—the problem of induction—is that our tendency to project past regularities into the future is not supported by reason. We are not justified in assuming that the future will be like the past. Sextus Empiricus is a precursor of this idea, writing in his *Outlines of Pyrrhonism* that if we focus on specific instances, "some of the particulars omitted in the induction may contravene the universal," while reviewing all of the particulars is "impossible, since the particulars are infinite"; therefore, "induction is invalidated."[24] That is, it occurred to Sextus and Hume that we might be turkeys.

For Hume, the best we can aspire to when predicting the future is relying on habits, on repetition. Since the sun has risen every morning that we've been alive, we are confident that it will rise again tomorrow.[25] The force of habit grounds our trust that the future will look like the past.

Habits are wondrous. I almost wrote my doctoral thesis on moral habits, before privacy seduced my curiosity. Habits allow us to live efficiently in the comfort of an illusion of safety. But relying too heavily on habits can be dangerous because the future doesn't always look like the past; if it did, no one would ever die.

The turkey is fatally unaware that Thanksgiving is just about to shatter its probabilistic worldview by an unexpected event that renders all previous observations irrelevant. The turkey might've been thankful for some philosophy. Don't be a turkey. Read on.

The Statistics Wars: Frequentists Versus Bayesians

I have a coin in my hand. It's a fair coin, in that it's evenly weighted. When I say that "there is a 50 percent chance of it landing heads if I flip it," am I making a statement about the coin or about my knowledge? Frequentists think that the probability of the coin landing on heads is a property of the coin. Bayesians think it's a statement about our degree of certainty.[26]

Frequentist statistics, going all the way back to Cardano, views probability as a long-run frequency of events. Probability is an objective property of the physical world, defined by the proportion of times an event occurs in a large number of repeated trials. Frequentist methods focus on sampling distributions and rely on concepts like hypothesis testing, confidence intervals, and *p* values. The fundamental assumption is that there is an objective truth that can be approached through repeated sampling and statistical analyses.

Bayesian statistics views probability as a measure of subjective belief or uncertainty. The Bayesian goes around collecting more and more data, updating her beliefs again and again, gradually reducing her uncertainty, but never achieving complete certainty.

Curiously, Richard Price gave a religious reason for publishing Bayes's paper. Price thought that Bayes's theorem aimed to show that "the world must be the effect of the wisdom and power of an intelligent cause; and thus to confirm . . . from final causes . . . the existence of the Deity."[27] Bayes's theorem has done many things, from cracking the Enigma code and helping win World War II to powering machine learning, but it hasn't proven the existence of God. Ironically, it took us away from causes, and away from religious thinking. Life is funny that way; even the destiny of ideas is unpredictable.

Frequentist and Bayesian statistics have opposite methods. Bayes's theorem asks how likely a hypothesis is to be true, given the data you've seen. Frequentists ask, if your hypothesis is true, how likely are you to see data as extreme as the one you have?[28]

Although debates between these two strands of statistical thinking can be very heated, one way to approach the divide, at least methodologically, is to think of the approaches as answering different kinds of questions and having different weaknesses and strengths. In a way, when we say that there is a 50 percent chance that our coin will land heads, we are saying something *both* about the world and about our confidence in our belief.

One of the strengths of Bayesian statistics is that you can consider all the data you have access to. If you do a study with frequentist statistics, you start from scratch; every new study is viewed in isolation, not taking into account data from previous studies. As is often the case, the strength of Bayesian thinking, taking into account prior data and updating it, is also its weakness: the choice of prior distribution. In Bayesian statistics, the selection of prior probabilities can significantly influence results, potentially introducing biases.

In 2023, a group of nine "expert forecasters" estimated a 30.5 percent probability that an AI catastrophe kills the majority of humanity by 2200.[29] In his book *The Art of Uncertainty*, David Spiegelhalter retorts that he is "not convinced these probabilities are much more than an expression of concern."[30] I agree. Only two pages later, Spiegelhalter writes that "we should endeavor to express our uncertainty through numbers whenever possible." I'm not sure putting a number, and one as high as 30.5 percent, to existential risk from AI is helpful at all. By putting a number on something, we make it look closer to truth than it is.

Another study, involving eighty-nine "superforecasters" (people who have been shown to be better than the average at forecasting), sixty-five domain experts, and fifteen experts in "extinction risk," came to very different conclusions. Superforecasters thought there was a 2.1 percent chance of an AI-caused catastrophe (killing at least 10 percent of the population), and a 0.38 percent chance of an AI-caused extinction by the end of the century; AI experts assigned the two events a 12 percent and 3 percent chance, respectively.[31] My own assessment is that the superforecasters are probably closer to an accurate prediction. But the point is this: Predictions are all over the place because we are clueless, which suggests those predictions are not saying much about the world.

The distinction between frequentist interpretations of probability and Bayesian ones matters for all kinds of reasons, but in this context here's one takeaway: When you read a percentage, look twice and ask yourself, Is this a prediction? If yes, put it in the "not-a-fact" box, and further ask yourself, Is it expressing something about the world, or is it expressing something about our ignorance?

The case of AI risk is a perfect example of numbers pretty much pulled out of thin air because we have arguably no good data to go on, and questionable common sense. When someone makes a prediction, depending on the context and their methods, as well as their will to power (what they want), they might be doing very different things: expressing their degree of confidence about something, expressing the chances of something happening in the world, expressing what they want the world to look like in the future, trying to manipulate their audience into buying what they sell, or a combination. They might not even understand what they are doing. What they are *not* doing, however, is uttering a true statement about the world, no matter which way you look at it.

Whatever your favorite interpretation of probability is, any prediction should be taken as an acknowledgment of uncertainty, whether the uncertainty lies in our beliefs, in the world, or both. We should be all the more sober about trusting predictions when considering just how chancy the world seems to be.

CHANCY ALL THE WAY DOWN

The Chancy Non-Laws of Chance

Charles Sanders Peirce and Friedrich Nietzsche were skeptical of the view of the world as a clockwork machinery. They argued that probabilities shouldn't be thought of as laws, that the order we see in the world is a kind of realization, a pattern that we find, or rather, project, in hindsight.

Peirce wanted to be an academic, but he failed to secure a long-standing job at a university. Instead, he worked for decades for the U.S. government, improving measuring devices in the Coast Survey. One of his most impressive achievements was measuring wavelengths of light through the length of a rod, thus tying the meter to an absolute standard of light.[32]

Peirce was also the first to use an artificial randomizer to choose the sequence of empirical trials. He introduced sampling and randomization in the design of experiments as a technique to draw statistical inferences. Peirce taught us that induction is a way not only of thinking but of doing.[33] Scientists manipulate their environment to draw more precise inductive conclusions than would otherwise be possible.

For Peirce, the universe is chancy all the way down, constants being variables that could have been otherwise. Denying determinism, he argued that reality is irreducibly stochastic; it can be analyzed statistically in hindsight but cannot be forecast with precision.

Peirce was reputedly solitary and cantankerous, but he recognized the role of community in the pursuit of truth: "Social sentiment [is] presupposed in reasoning." Contrary to Descartes's idea that our thinking is solidly founded on the individual, for Peirce it's society who ultimately decides what counts as true. That's why it matters that we recognize that predictions are not facts; if we treat predictions as facts, if we count them as truths, then we are bound toward injustice and a loss of freedom, as we will see in the following chapters.

Nietzsche agreed with Peirce. According to him, everything in the world, including laws, is the product of chance. What Nietzsche calls "necessity," or destiny, is just another throw of the dice, another product of chance. In the infinite throw of dice that is life, there are throws that resemble purposiveness and rationality.[34] Life is a gambling game. Foolish players play the odds as if their life were one among a large number of games or throws of the dice.[35] The smart players know that the life they have is one and only one; one opportunity that came to be among many others that didn't; one chance to connect with others, to enjoy, to take it all in and contribute to the world.

Even Buddhists, who believe in reincarnation, think that a human life is as precious and rare as a blind turtle emerging from the water once every hundred years and being lucky enough to surface at just the right time and place to put its head through a golden yoke floating in the vastness of the ocean. I haven't seen many golden yokes floating around at sea.

Of Cats and Butterflies

Two scientific developments support the idea that the universe is chancy all the way down: quantum mechanics and chaos theory.

Quantum mechanics is a fundamental theory in physics that describes the nature of matter and energy at the atomic and subatomic scales. It reveals that particles can exist in multiple states simultaneously and are governed by probability rather than definitive trajectories.

Werner Heisenberg's uncertainty principle states that we cannot simultaneously know both the precise position and the momentum of a particle. The constraint isn't merely a limitation of measurement but a statement about the nature of reality itself; particles exist in a probabilistic haze, their exact state indeterminate until observation collapses their wave function into a singular manifestation.

Schrödinger's cat is a famous thought experiment in quantum

mechanics designed to illustrate the paradoxical nature of quantum superposition. In this hypothetical scenario, a cat is placed in a sealed box with a flask of poison and a radioactive source. If a random quantum event occurs (like a radioactive atom decaying), a mechanism will break the flask, killing the cat. According to quantum mechanics, until the box is opened and an observation is made, the cat exists in a simultaneous state of being both alive and dead—a superposition of states.

This thought experiment was proposed by the physicist and Nobel Prize laureate Erwin Schrödinger in 1935, while he happened to be living in a redbrick house on 24 Northmoor Road, a leafy quiet street in Oxford. Two doors south, in number 20, lived the author of *The Lord of the Rings,* J. R. R. Tolkien. Unfortunately, a fact rarely mentioned is that Schrödinger was also a pedophile who kept a record of his abuses in his personal diaries, where he also justified his actions by arguing that the innocence of girls was a perfect match for his genius, "since no woman will ever approach nearer to genius by intellectual education."[36]

The cat example was meant to highlight the absurdity of applying quantum mechanical principles to macroscopic objects. The lesson for the purposes of this book is that whether the cat will be alive or dead when we open the box is entirely unpredictable.

A skeptic might argue, Okay, fine, quantum mechanics suggests that uncertainty is part of the fabric of subatomic particles, but that doesn't necessarily show it's relevant to our macroscopic world of things. That's a fair criticism, and precisely the point of the thought experiment: The small uncertainties about the particles that make up objects follow the law of large numbers and average out.[37]

Quantum mechanics is worth bringing up, however, first, because it shook scientists' and the public's confidence in a deterministic clockwork model of the universe. Second, it's a powerful metaphor for the nature of reality, and metaphors can be good teachers. To better understand unpredictability in our macroscopic world, it's more appropriate to look to chaos theory.

The mathematician and meteorologist Edward Lorenz was researching weather forecasts in the 1960s. He was simulating the weather with a model of the Earth's atmosphere. The computer would print out the results of the model as a series of numbers representing the value of each of twelve variables (wind speed, air pressure, humidity, and so on). To save paper, it printed the numbers rounded to three decimal places, instead of the more accurate six decimal places the computer was working with.

When Lorenz repeated one of his simulations with the printed numbers, it produced a dramatically different weather forecast. It was just a fluke that he repeated the simulation with the numbers rounded to three decimal places rather than six, but it led to a whole new theory.

Against the scientific consensus of the time, Lorenz argued that minute differences in initial conditions could translate into dramatic changes in the final conditions. He shared his theory in a paper wonderfully titled "Predictability: Does the Flap of a Butterfly's Wings in Brazil Set Off a Tornado in Texas?"[38]

Chaos theory adds another layer to our cosmic uncertainty, showing how even deterministic systems can produce wildly unpredictable outcomes. A minute change in initial conditions can exponentially transform a system's trajectory, rendering long-term prediction impossible.

Together, quantum mechanics and chaos theory paint a radical portrait of existence: a universe not of fixed, mechanistic certainties but of fluid potentialities and intricate, interconnected probabilities. They suggest that uncertainty is a fundamental characteristic of reality, and statements about the future, guesses.

PREDICTION AND JUSTICE

The problem of likeness to truth presents challenges for justice. Predictions have moral weight when they have an impact on people's well-being; they are related to justice when they are involved in granting people what they deserve.

People are being denied opportunities—loans, jobs, grants—as a result of what other people think the future will look like. Prediction goes beyond standard theories of distributive justice because resources are being allocated not only with respect to present and past circumstances—who people are, and what they deserve given what they have already done—but on the basis of a guess as to who people will become. Is that fair? Shouldn't fairness be more closely related to merit?

Reminding ourselves that predictions are not facts matters for practical purposes. It limits possible injustices.

Confusing predictions with truth is dangerous not only because we are more likely to acquire false beliefs, which can lead us to make mistakes as lawyers or business owners or citizens, but also because it can have a negative effect on how we build our future. That leads us to the topic of the next chapter: Predictive societies tend toward oppression through surveillance and self-fulfilling prophecies because predictions are speech acts that are closer to commands than to truth claims.

CHAPTER FIVE

WHEN PREDICTIONS BECOME VERDICTS

*On the tyranny of surveillance and
self-fulfilling prophecies*

The Roman emperor Caracalla was born and killed by prophecy. His mother, the Syrian Julia Domna, had been predicted to marry a king. His father, the emperor Septimius Severus, a highly superstitious man, had been looking for a wife. Keen to find favor with fortune, he had sought a woman whose horoscope foretold that she would marry a king. Julia and Severus married in 187. Was Julia's fate written in the stars, or was the astrologer's prophecy its ultimate cause?

Thirty years later, one hot spring morning in the eastern provinces of Rome, a messenger delivered the latest imperial dispatches to the now emperor Caracalla. The emperor was on his way to a chariot race, and he asked his adviser and Praetorian prefect, Macrinus, to read the letters for him and report back on anything important. In the correspondence, Macrinus found his own death sentence in the form of an imperial prophecy.

Caracalla, superstitious like his father, and with a reputation for cruelty, had requested that a prophet ask the gods about any possible rivals who might wish to overthrow him. The letter recounted how the soothsayer had foreseen that it was Macrinus who was destined to imperial power. The rumor was spreading through the empire. When it inevitably reached the ears of the despot, Caracalla would undoubtedly order the execution of his prefect. Macrinus had no time to lose. He hired Martialis, an aggrieved soldier, to save his life by murdering the emperor. When Caracalla was relieving himself off the side of the road during a trip, Martialis crept up behind him and stabbed him in the back with a dagger. And so it was that the tyrant fell. Macrinus ended up succeeding him, exactly as the prophet had foretold.[1] Would Macrinus have become emperor without a prophecy foretelling his reign?

There are a few ways others can know what you'll do in the future.

One method is through a self-fulfilling prophecy, a prediction that causes itself to become true through changing people's beliefs and actions. A prediction about people has reflexive qualities, acting like a magnet that bends reality toward itself. When the force of the magnet is strong enough, the prediction becomes the cause of its becoming true. Clouds and rain are impervious to weather predictions, but people's fates are partly shaped through expectations.

According to Pygmalion theory—named after the mythological sculptor who fell in love with a statue he made—when leaders have high expectations of others, students do better in education, and employees perform better in business.[2] If I'm your supervisor, you ask me for a recommendation letter, and I write in it that I don't expect you to have an academic career, no one will hire you. My prediction that you won't have an academic career almost guarantees you won't have one. Self-fulfilling prophecies also affect collective behavior. When important institutions like the Federal Reserve predict interest rates going up, it has an effect on how people behave, which in turn affects interest rates; borrowers rush to secure loans at current lower rates, savers delay purchases, investors sell stocks and buy bonds, all of which affects the supply and demand for credit, with an increase in borrowing pushing rates up.

A second method to forecast someone's future is to alter the world in a way that makes your prediction come true. Macrinus knew Caracalla would fall because he ordered his assassination. Hannah Arendt argues that totalitarian leaders' concern to make their predictions come true overrules all other considerations; at the end of the war, the Nazis were willing to use their force to destroy their country "in order to make true their prediction that the German people would be ruined in case of defeat."[3]

A third method is through extensive surveillance. You could produce a very smooth, shiny phone that instantly connects people to their loved ones and provides countless distractions. It also happens to be a spy in their pocket that collects *thousands* of data points per day about them: everywhere they go, everything they search for, every-thing they watch, everything they read, every person they interact with. If you have data on someone since they had their first phone and you know they've woken up every day at 7:00 a.m. for the past decade, it doesn't take psychic powers to know at what time they'll wake up tomorrow.

What is common to these three methods of prediction is that they are despotic. At best, by making a prediction about you, I take power away from you, however justified the end might be. At worst, they can turn into egregious abuses of power. The main promise of prediction is the enhancement of power. It is fitting that its main peril is the abuse of power, epitomized by a descent into authoritarian tenden-cies. The more predictions alter the situation they predict, the more they become rulings, and the more scope there is for abuses of power.

Machine learning systems are experts at using all three methods: Their predictions often become self-fulfilling prophecies; they are modifying the world to fit their image, much as statistics did in the nineteenth century; and they are fed through a network of mass sur-veillance, which in turn they facilitate by being capable of quickly sifting and analyzing data.

Here's the upshot of this chapter: Predictions are commands dis-guised as descriptions, and the more we allow companies and govern-ments to use prediction across society, the more our future is being decided by them, because the most effective way to predict the future is to determine it. Prophecies are not innocuous. Predictions change the reality they purport to predict, and sometimes their accuracy is evidence not of knowledge or understanding but of despotism. When we predict the future of individuals' lives as if they were the weather, we are treating them like things and not like persons.

When we obey predictions, when we believe them unreflectively and act in accordance, predictions become rulings. A prediction can be as definitive as a death sentence. When self-fulfilling predictions are isolated occurrences, they can lead to unfairness. When they become systemic, they lead to catastrophes, including financial crises and tyranny.

This chapter follows a trail. It starts with a philosophical exploration of how to do things with words. I then go on to argue that predictions can only give power to prophets if we believe them. The chapter ends with an exploration of self-fulfilling prophecies on the ground: in the lab, in business, in society at large, in the context of AI, as perfect crimes, as products of surveillance, and in oppressive systems.

HOW TO DO THINGS WITH PREDICTIONS

J. L. Austin thought that words do much more than describe the world. He was a British philosopher who acted as an intelligence officer during the Second World War, and then became professor of moral philosophy at Oxford. In 1955, he delivered a series of lectures at Harvard that would later be published under the title *How to Do Things with Words*.

Austin argues that there are some sentences—*speech acts* or *illocutionary acts*—that are neither true nor false, but rather performative utterances that do not describe the world, and therefore have no truth value. To utter a speech act amounts not only to "saying" something but also to *performing* some kind of action. When a city official says "I pronounce you husband and wife," she is marrying a couple. When your friend says "I promise I won't tell anyone," he is committing to doing (or, rather, not doing) something. When your mother writes "I bequeath my house to you," she is leaving you her property. And when a naval officer says "I name this ship *Prophecy*," she is christening a ship. These are all speech acts. To determine whether something is a speech act, it is necessary to analyze both the words and the context.

Predictions are speech acts. More specifically, predictions are misleading speech acts. When someone marries you with a sentence, there is no deception involved; it is clear what the sentence is supposed to achieve. In contrast, when a financial pundit predicts that a country is going to have a financial crash, it seems that they are merely describing the future, when in fact they are instigating investors to flee that country. When someone makes a prediction about human affairs, they seem to be describing the future, but they are influencing it—even when it is not their intention.

"The prophet's dilemma" arises in cases of predictions that are warnings of disasters to avoid. When an epidemiologist says that a pandemic will cause a million deaths if we don't isolate, the prescriptive action is to quarantine. If people believe the prediction and quarantine, and there are no deaths, the person can be accused of a false prophecy when in fact it was a self-defeating prophecy.

Predictions about human beings attempt to change the future by altering what people believe and how they behave, which is why they are veiled imperatives or orders. When the CEO of a tech company says "In the future, everyone will use AI," he is trying to bend reality toward that vision; in a way, he is saying something like "Go forth and get your AI before you fall behind! Go forth and fulfill my vision!" His prophecy is therefore not descriptive, but *prescriptive,* in that he is proposing a certain course of action. People who listen to that prophecy and act as if it were true, say, by buying the AI for fear of missing out, might think that what they are doing is getting ahead of the curve, but what they are actually doing is akin to obeying an order.

DO NOT OBEY IN ADVANCE

"Anticipatory obedience is a political tragedy," writes the historian Timothy Snyder. It's his first lesson in resisting tyranny. Obeying in advance means adapting to a situation instinctively, unreflectively, before you receive explicit orders.

An analysis of predictions as speech acts partly explains why and how we tend to acquiesce to authoritarian leaders: by believing and acting in accordance with their predictions. We don't even notice we're obeying. Perhaps because the structure of sentences that express predictions is so similar to those that describe the world, if we are not mindful, our psychology tends to accept other people's visions as inevitable truths that are on their way to materializing. But if we take a second to recognize them as predictions, and if we understand that predictions are prescriptive and not descriptive, we stand a better chance of controlling our fate.

INVITATIONS FOR DEFIANCE

We can take predictions as inevitable prophecies or likely outcomes and comply with them, or we can interpret them as an invitation for defiance. Sometimes predictions are so close to commands that they amount to rulings. But we should remember we still have the choice to disobey. How we will react to a prophecy depends on how much credence we give it. Joe Frazier didn't take Muhammad Ali's prediction that he'd win at face value; it made him angrier, and he ended up winning the fight.

In the case of an epidemiologist with a good track record and who stands to gain nothing from averting deaths (other than the satisfaction that entails), we have more reason to lend our credence to her prediction. In the case of a tech company executive with an appalling track record of lies and exploitation who could not have more of a financial interest at stake, not so much.

Hannah Arendt makes the case that debating "about the truth or falsity of a totalitarian dictator's prediction is as weird as arguing with a potential murderer about whether his future victim is dead or alive." The prophet can kill his victim to prove his statement right. The only appropriate response is therefore to "rescue the person whose death is predicted."[4]

FATE OR FAITH?

Forrest and I are sitting across from each other having dinner. His face turns serious when I tell him about this chapter. "When she was very young, my wife got told by a psychic who read her palm that she'd live a short life," he says. She died a few years ago, young, from cancer. Sadness flashes across his face. How dare a psychic forecast an untimely death to someone as suggestible as any young person is likely to be.

Many years ago, I went through a phase in which I got obsessed with death, reading every book on the topic that crossed my path. My high school teacher and mentor, Roger Gouran, the person who had taught me Latin and Shakespeare and to whom I owe an education, had died recently, and a few other incidents had made me anxious about my own mortality. Would I live long enough to read half the books on my shelves, to write the books I have in me?

My brother Ivan had gone to a Buddhist event in which one high lama gave him a clay tablet made to grant a long life. Aware of my anxiety, Ivan gifted it to me, wrapped in a white silk scarf, along with a mantra on a laminated blue card. And that was the beginning of the end of that phase—as if my big brother could command the universe to protect me. I can honestly say that I don't think I'd live a day less if I were to lose any of these objects, but I've carried the mantra in my wallet every single day for a decade and a half, and the clay tablet is safely stored in a box, along with my mentor's favorite handkerchief and a few other quirky treasures. We are suggestible creatures.

Lucky charms are embodied predictions. They make us feel safe because we explicitly or implicitly attribute to them the power of commanding that the world intercede in our favor.

Self-fulfilling prophecies are as old as predictions. Perhaps the most paradigmatic one is that of Oedipus. Had it not been for him learning about the prophecy that he would kill his father and marry his mother, Oedipus wouldn't have tried to flee the people he thought were his parents, and he wouldn't have fulfilled the soothsayer's vision.

The first lesson to draw from self-fulfilling prophecies is that faith in prediction matters. Predictions can only cause themselves to become true if we believe in them. If the emperor Septimius Severus hadn't trusted fortune tellers, he wouldn't have searched for a woman who was forecast to marry an emperor. If Oedipus had reacted with laughter instead of terror at his fateful prophecy, he might have never left Thebes.

Would you find it easy to dismiss the prediction of a psychic? Maybe. But I wouldn't count on it. Human beings tend to be superstitious. It's a secondary effect of our being good at finding patterns. We can't help seeing possible causal relationships between almost any two data points, even when we are confident there's nothing there.

I noticed that whenever my brother July missed a point in tennis, he'd grumpily stick the offending ball into his pocket and refuse to play with it until another offending ball took its place. "Are you punishing the balls?" I asked, amused. He nodded, laughing sheepishly.

It's all the funnier because he is perhaps the most rational person I've met.

In July's defense, Michael Jordan consistently wore his University of North Carolina shorts under his Chicago Bulls uniform for good luck, Serena Williams would not change her socks during a tournament run if she was winning, and I'll spare you Rafa Nadal's *dozens* of superstitious rituals; watching his routine is quite the spectacle.

The better we understand predictions, the more we demystify them, and the less power they have over us. Even if we are naturally superstitious, it can still be helpful to rationally lower our credence in predictions—especially when they are issued by dubious types with financial interests.

I recently met a leading filmmaker at drinks. He was telling me about a documentary on AI he had made. "We interviewed everyone," he said, before reciting a litany of names of tech executives, followed by grandiose views about how AI will change our future. "Have you interviewed anyone without a financial interest in AI?" I asked. "You're asking the right questions," he said, before walking away.

SELF-FULFILLING PROPHECIES IN THE LAB

Self-fulfilling prophecies are a well-known effect documented by science.

In 1963, Robert Rosenthal and Kermit Fode asked a group of psychology students to run rats through a maze learning task. The experimenters told half of the students that their rats were specially bred to be "maze-bright" (intelligent) while telling the other half that their rats were "maze-dull."

Students weren't surprised when the smart rats did better at finding their way through the maze than the dull ones. The researchers, on the contrary, were astounded, because in fact all of the rats were regular, randomly assigned to students.[5]

The rats that were cared for by students who thought them smart learned the mazes more quickly. Without realizing, students had treated "bright" rats with care and patience, and "stupid" ones with contempt. They got what they expected, because they involuntarily acted in accordance with those expectations; they obeyed the researchers. When the researchers told the students that their rats were smart, they were doing more with those words than conveying (false) information; they were helping create a reality.

Another known phenomenon in the lab is the *placebo effect*, whereby a person's belief in a treatment can produce physiological improvements in their condition, even when the treatment contains no active therapeutic ingredients. This effect has been documented across a wide range of conditions, from pain management to depression, and can trigger measurable changes in brain chemistry, immune response, and hormone levels.

The *nocebo effect* is the negative counterpart to the placebo. It's when a person's pessimistic expectations or beliefs about a treatment or situation lead to harmful or unpleasant effects, even in the absence of any harmful stimulus. The nocebo effect can manifest itself in many ways, from experiencing side effects from sugar pills, to feeling pain or discomfort from a procedure because it was anticipated. For example, studies have shown that when patients are told a procedure might be painful, they often report higher levels of pain than those who weren't given such warnings.

SELF-FULFILLING PREDICTIONS IN BUSINESS

Books

I've never visited a more vibrant city than Madrid, and it just so happens that my Spanish editor, Miguel Aguilar, is in town. He suggests we meet at a cocktail bar. It's still light out, and at 5:00 p.m. the waiter lowers a curtain to create a twilight effect. I squirm in my seat, thirsty for every possible ray of sun I can take in. I'm just about to beg the waiter to pull up the curtain when Miguel enters the bar, his salt-and-pepper beard framing an easy smile. He is beaming because one of his authors, Han Kang, has just won the Nobel Prize.

"Do you know why I jumped at the chance of *Prophecy*?" he asks. I shake my head, relishing the attention and compliment while I sip on my *mezcalita*. "Because I have a front-row seat to how predictions are shaping our literary world. I can guess how what I'm seeing is happening in every sphere of business, and no one is calling it out."

Self-fulfilling prophecies are part and parcel of the publishing industry. Every author hopes for a large book advance. It's not (only) about the money. A large advance is a self-fulfilling prophecy: The authors expected to attract attention and resources from readers (in the form of book sales) are given the most attention and resources by their publishers (in the form of publicity), which in turn translates into getting more readers.[6]

You want your editor to be as excited about your book as you are, and willing to go out there and fight in its corner. A TV presenter once asked the famous editor Robert Gottlieb whether he'd ever had the experience as a publisher of rejecting a manuscript that went on to be very successful. "Oh, everybody's had that experience," he said, to which the presenter went on to ask if that was a nightmare. "No, because if you're sensible, you remember that you turned it down because you didn't much like it, and that if you had published it not very much liking it, it probably would not have become a great success."[7]

One way to show enthusiasm for a book is with a big advance. Very few books for which publishing houses have paid a million dollars or more fail. In some cases, that's partly because these are good books that deserve the huge readership they command. More often than not, authors of million-dollar books are megastars, sometimes for good reasons, other times for terrible ones. If a person has enough fans who read, their book will do well regardless of whether it's good. Just as important, publishing houses try to make sure these projects don't fail.

If a book with a low advance fails, it's sad for the author, but the publishing house will survive. If a few of the most expensive books fail, however, a publishing house can be in financial trouble. That means that publishers will do everything they can to make their big projects sell well by pushing them forward with all the firepower of their publicity and marketing teams. Prospective best-selling authors tend to get a disproportionate share of the houses' focus.[8] Expensive books, it can be argued, are a kind of manufactured success.

Some editors aren't too confident in the power of self-fulfilling prophecies, because books are (thankfully) unpredictable. "Every year a million-dollar book falls flat on its face, especially in fiction," one New York City editor tells me.

In the United States, there are about twelve hundred books every year that are branded "anticipated best-selling books" for which a high advance is paid; that's 2 percent of all books that get published in a year by commercial firms. Those twelve hundred books get jointly paid $1 billion. Book sales are highly correlated with the level of advances paid. Of the twelve hundred top-paid books, a high proportion of those are paid advances of $1 million or more. Let's say six hundred or so books are bought for $1 million or more. If every year one or two of them flops, that's a 0.3 percent rate of failure.[9] The self-fulfilling prophecy seems near perfect.

However, it's hard to causally tease apart the effect of the prediction itself from the highly correlated elements of having a large following and having published a best-selling book before. Because these prophecies are based on historical data, they inevitably favor those who have been favored before. In nonfiction especially, the person more likely to get a high advance is someone who looks like Richard Dawkins. At the University of Oxford, 80 percent of professors are men, so that figures.[10] The self-fulfilling prophecies that favor some over others compound, further enhancing their effects.

But editors are right that books are marvelously unpredictable, and undervalued books are where most of the surprises lie. We'll leave those for the next chapter.

When There Is Little More Than Prophecies

In the case of expensive books, there is good reason to think that a self-fulfilling prophecy will work its magic: There is an author who is already well known, there is a book that a competent editor has vouched for, and there is a publicity and marketing team ready to catapult it.

But sometimes businesspeople think that predictions are all it takes. Outside religion and sports, business is the field in which I've found most instances of magical thinking and faith in prophecies. The business culture often mixes a host of approaches: the rational (a good idea), the emotional (gut feelings or fear of missing out), the deceptive (fake it until you make it), the wishful thinking (too much faith in the power of "manifestation," or AI for that matter), the superstitious (signs or portents), and the devotional (business gurus), among others. The latter four make businesspeople particularly vulnerable to prophecies.

One hypothesis is that the business context can be so stressful, with so much at stake, that engaging in magical thinking alleviates anxiety by enhancing self-assurance.[11] Having a confidence boost is a good thing as long as there is good reason to feel optimistic; otherwise, it can make the inevitable crash against the wall all the more painful and serious. Two of the most salient cases in the past few years of confident leaders who had little reason to feel assured are those of the Fyre Festival and Theranos.

Fyre was a luxury music festival planned for 2017 in the Bahamas. It was heavily promoted by influencers but completely fell apart when attendees arrived to find emergency disaster tents instead of the promised luxury accommodations, inadequate food and water, and none of the entertainment or amenities advertised. The festival's organizer, Billy McFarland, was later sentenced to six years in prison for fraud and ordered to forfeit $26 million.

Theranos, founded by Elizabeth Holmes, was a biotech start-up that claimed to have revolutionary blood-testing technology requiring only tiny amounts of blood, but investigations revealed that the technology didn't work and that the company had been deceiving investors and patients, leading to Holmes being convicted of fraud in 2022 and sentenced to more than eleven years in prison. The company, once valued at $9 billion, collapsed in 2018 after a series of *Wall Street Journal* exposés.

Both founders, Billy McFarland and Elizabeth Holmes, didn't seem to start out with an *intention* to defraud people. They were faking it in the hopes of making it, but instead of focusing on doing a good job, on building a festival or innovating in blood tests, they focused on positive thinking and predicting bright futures.

Silicon Valley is a great example of why we shouldn't give in to the allure of bright but baseless forecasts. "There is a relentless belief in the future here," Sam Altman, CEO of OpenAI, once said. "There are people here who will take your wild ideas seriously instead of mocking you."[12] That is perhaps Silicon Valley's greatest virtue and vice. The tech culture incentivizes entrepreneurs to engage in extreme wishful thinking, as Parmy Olson expertly describes in her book *Supremacy*.

The person who ends up getting funding from venture capitalists is not the sober realist who is thinking not only of the best possible case but also of what can go wrong and how to minimize risks, but the dreamer who paints the most optimistic yet compelling picture for investors. And what men in Silicon Valley see as a bright future (for them) can sound like a dystopia for the rest of us. Zuckerberg might feel excited at the prospect of people wearing the glasses he's selling, which can record video and sound and take people's attention away from the present moment with notifications, but it is not the world I wish for myself or my loved ones. Zuckerberg is likely thinking about the money and power he'd accrue, while I'm concerned about the autonomy and freedom we'd lose.

It is because of that culture that incentivizes exaggerated prophecies that we end up with cases like Fyre and Theranos. Although these are examples of failed self-fulfilling prophecies, the prophecies were successful enough to create expectations that caused harm. People who wanted to attend Fyre paid thousands of dollars to experience a nightmare. There were Bahamians who worked for weeks for the festival without getting paid. Theranos gave people false medical results. Investors in both projects lost millions.

Prophecies are dangerous because they invite us to change our expectations, even when there is no good reason to do so. If we were more conscientious and remembered Aristotle's lesson that predictions convey no new information about the world, we'd stand a better chance of not being as susceptible to them. Building resistance against prophecies is vital to avoid them having systemic reach.

SYSTEMIC SELF-FULFILLING PROPHECIES

The Great Depression and Bank Runs

Winston Churchill, who had a knack for finding himself in the thick of things, happened to be visiting New York City on October 24, 1929. He stayed at the Savoy-Plaza Hotel while he took care of business arrangements and literary contracts. "Under my window, a gentleman cast himself down fifteen storeys and was dashed to pieces, causing a wild commotion and the arrival of the fire brigade," he wrote.[13]

Afraid of bigger losses, traders started selling stocks. Selling created a feedback loop driven by panic: As more people sold their stocks, prices fell further, which confirmed investors' fears and prompted even more selling. Soon, prices plummeted and stocks were selling for next to nothing. Crowds gathered outside Wall Street, where more than twelve million shares were sold in a day. People in the streets of New York's Financial District were in tears, the sidewalks filling with shell-shocked traders and ruined investors. Terror gripped the city.

The crash became a self-fulfilling prophecy because the very act of trying to prevent losses by selling stocks created the massive losses that investors feared. The market's decline wasn't necessarily inevitable based on economic fundamentals alone; it was collective fear in action. Panic spread to the broader economy as people lost confidence in banks and began withdrawing their deposits, further weakening the financial system.

The phrase "self-fulfilling prophecy" was coined by the sociologist Robert Merton, who used bank runs as an example.[14] Even the most solvent banks are vulnerable to a self-fulfilling prophecy, because they keep only a fraction of depositors' money readily available as cash while lending out or investing the rest. If they kept all our money in cash, then they wouldn't have a business model. If enough clients ask for their money back at the same time, the bank is in trouble. That's what insurance is for, you might think, and that's true, but if enough of the population wants their money back across the country, then insurance companies will be in trouble too.

Wars

Thucydides, the great Athenian historian and general, wrote in his account of the Peloponnesian War, "What made war inevitable was the growth of Athenian power and the fear which this caused in Sparta."[15] A proud and rising Athens expected greater respect. The Spartans thought the Athenians arrogant, ungrateful for not recognizing the role that Sparta played in maintaining the order that provided the stability Athens had needed to thrive.

What is now dubbed Thucydides's Trap refers to fearful responses to shifts in the balance of power that in turn lead to conflict. The First World War has been interpreted under such a framework. At the beginning of the twentieth century, Germany became the dominant military and economic power in Europe. It sought colonial expansion. German exports rivaled and threatened British trade networks. When Germany's share of global manufacturing surpassed that of Britain, tensions rose. Germany built up its navy, which led to an arms race that eventually broke out in war. Contemporary relations between China and the United States, with escalating tariffs and sanctions since 2018, could be another example.[16] The more we fear a conflict, the more we prepare for it, and the more we make it likely that it'll happen.

Racism and Sexism

Self-fulfilling prophecies are to racism and sexism what kerosene is to fire. In his pioneering paper, Merton gave the example of white workers not wanting to hire Black workers on account of an expectation that they would be strikebreakers because they weren't part of the tradition of unionism and collective bargaining. But by excluding Black people from jobs and unions, the unionists were forcing Black workers to be strikebreakers.

In *The Authority Gap,* Mary Ann Sieghart documents how women are often not taken as seriously as men. Women are constantly interrupted when trying to voice their opinion, subjected to everything from being ignored to being harassed, blamed for their shortcomings if they act submissively, shamed for being aggressive if they act assertively, and constantly held to higher standards than men.

In the context of venture capital, companies founded by women receive less than 3 percent of all investments, and only between 10 and 15 percent of investors are women. To try to balance the situation, some women are investing more heavily in women. However, research suggests that receiving funding from women only makes it harder for female founders to raise additional rounds of financing. When investors see that a woman received funding from a male, they assume competence, but when the funding comes from women, they assume it's because of her gender.[17] Women need funding from men to thrive in a male-dominated context, but men won't fund women for the sole reason that they're not men.

Minorities fared even worse. Latino founders are also discriminated against, with only 2 percent of venture capital going to them.[18] Black founders got only 0.48 percent of venture dollars in 2023.[19]

Jean Brownhill is the founder and CEO of Sweeten, a company that helps homeowners match with general contractors. She managed to raise $8 million, but only after pitching the idea to 250 investors. One investor who had loved the idea still turned her down. "As a Black woman you're going to have a harder time fundraising, a harder time retaining talent, a harder time selling," the investor said.[20]

No matter how brilliant and capable a Black woman is, she might not find anyone to support her on the pretense that others won't support her either, thereby creating a self-fulfilling prophecy. We are losing out on talent and great ideas. We are building systems in which you are almost guaranteed failure irrespective of merit for not being a white man. If Silicon Valley and VCs were half as contrarian as they boast to be, they'd be eager to fund those who are discriminated against by the rest of society.

Police and prosecutors sometimes use the same excuse not to charge rapists. If they think the jury might not convict the perpetrator because of sexist tropes, the police or prosecutors might not bother charging him, thereby creating a self-fulfilling prophecy that makes it impossible to overcome the sexist tropes that result in violence against women and girls.[21]

Police then have the temerity to argue that *they* are not sexist, that it's juries who are sexist, even though juries almost never get the chance to prove the police wrong. That's self-fulfilling prophecies at their very worst, in the hands of authorities who are tasked with protecting victims but who end up protecting criminals.

Political Polls

"Good polls can reveal the will of the people," G. Elliott Morris, an American data journalist, wrote.[22] George Gallup, a professor of journalism who became director of research at an advertising agency, took what he had learned in market research and turned it into political polling with sample surveys in the 1930s. If it works for toothpaste, why not for politics, Gallup thought.

Pollsters described their surveys as "snapshots," a kind of X-ray of public opinions. Citizens no longer needed to vote to make themselves heard. "The will of the majority of citizens can be ascertained at all times," realizing a "truer democracy," ensuring that dictatorships would be a thing of the past. Again, this is the 1930s. Elmo Roper, another leading pioneer of political polls, agreed, arguing that polls were "the greatest contribution to democracy since the introduction of the secret ballot."[23]

Gallup compared his role to that of the meteorologist. Polls, from their very beginnings, were not only about learning about the present and what people think today; their primary function was looking to the future.

Political polls suffer from two main problems. First, there is no such thing as "public opinion." When people are forced to answer questions worded in one way or another which necessarily simplifies the matters at stake, and we get a percentage, the result is the illusion of knowledge. Public opinion is not a percentage. The second, more concerning problem is that polls affect people's opinions in various ways. In an experiment in 1980 and replicated in 1995, people were asked whether the 1975 Public Affairs Act should be repealed. A third of people gave an opinion, even though the act doesn't exist. A U.K. poll found that almost half of respondents had an opinion on a nonexistent, but purportedly quite popular, politician.[24] By asking a question—any question—you are already swaying individuals' opinions.

Furthermore, polling institutions are not neutral. At the very least, their livelihoods depend on coming up with engaging results. Polarizing and scandalous results will be better for their business—but not for democracy. Often, polling companies are partisan.

Gallup said that "election forecasting" was one of pollsters' "least important contributions," and Roper said they are "socially useless" and might "do very much more harm than good." However, election forecasts have always been their main product. Every election there are thousands of polls done, and there is no reason to think they are good for democracy.

Polls might work as well for toothpaste as for politics, but in the case of toothpaste there is a marketing team wanting to sell a product. If democracy is for sale, it isn't democracy. Polls distort democracy.

One hypothesis for the failure of polls to predict Trump's win in 2016 is that college graduates were overrepresented in the samples and more likely to vote for Clinton, but another hypothesis is that, thanks to polls, Democrats were too confident in Hillary's win to go and vote; Trump winning was unimaginable to them, so they stayed home.

In the 2022 elections, *The New York Times* suggested that skewed surveys "polluted polling averages." The "bazaar of polls" used opaque methodologies, and even two high school junior pollsters were being included on the influential statistical analysis website FiveThirty-Eight. The director of one firm placed bets on the election results, raising questions of possible conflicts of interest. "These frothy polls had a substantial, distorting impact on how people spent money—on campaign strategy, and on people's expectations going into the election," said Steven J. Law, the chief executive of the Republican super PAC the Senate Leadership Fund.[25]

In the 2024 American election, commentators pointed out that polls were used to influence the results of the election. Quantus Insights, a polling firm that supported Trump, took credit for the shift from Kamala Harris leading slightly to Trump marginally overtaking her. "You're welcome," posted the firm on social media.[26] For months, the edge of one candidate over the other was so slight as to be insignificant, but their supporters clung to the minor lead as if the election had been won, creating the impression of a more significant advantage than the data suggested.

Polls can either encourage people to vote or discourage them from doing so. Both effects are undesirable because they don't instill a sentiment of citizen duty that would encourage people to always vote according to their conscience. Polls can also push people to vote for someone they wouldn't have otherwise voted for. We are social animals; we tend to follow the crowd. In contrast to public debate, polls influence voters in the wrong ways, for the wrong reasons. Many elections are won by a narrow margin, and the data of what would've happened in a world without political polls doesn't exist. Might self-fulfilling prophecies be stealing elections?

SELF-FULFILLING PROPHECIES AS THE PERFECT "CRIMES"

When self-fulfilling prophecies work, they become the perfect "crime." By bending reality to their prediction, they erase all evidence of their influence. The genius of self-fulfilling prophecies is that they can turn something that wouldn't have happened into a fact, all the while giving the appearance of neutral descriptions of the world. It's like a murder weapon that makes itself and the body disappear when it strikes, or a computer virus that covers its own tracks.

Because medicine has to do with life and death, it is a context in which the danger of self-fulfilling prophecies is especially salient. Medical resources are scarce, and health-care professionals are forced to engage in triage: They have to prioritize patients according to the severity of their needs and the expected value of the care.[27] That means that if you go to a hospital, a doctor (or an algorithm, these days) will make a prediction about you that might decide whether you live or die. If there is another patient needing care who is just as sick as you, but has better prospects of surviving, maybe because he's younger, or his vitals are slightly better, they might provide care for him first. The forecast that you are likely to die can end up killing you.

Years ago, I met a paramedic who was transporting a (dead) organ donor to a hospital. The donor started having a spontaneous heartbeat. When the paramedic called his supervisor, he was reminded that that wasn't a patient but a donor. When someone is declared dead, there is an implicit prediction that that person is beyond possible help. But sometimes people whose hearts have stopped wake up again. Prophecies are prescriptive; they tell us how to act.

This supervisor will have to live with that call on their conscience, but most health professionals will never know whether they made the right choice. They might unconsciously defend the precision of forecasts because it's too painful to take responsibility for decisions that might spell someone's demise. Those patients deemed beyond hope will almost always die, and with them the data that could've told us whether they might've lived, had we prioritized them. Another man who had been labeled "organ donor" revived in the operating room just as his organs were going to be removed and went on to live for years.[28]

Except for very few cases, like the one in which the presumed dead person wakes up, self-fulfilling prophecies don't create error signals, thereby deflecting scrutiny and shrouding injustice.[29] That makes them a threat to impartiality and accountability. The deception and despotism of self-fulfilling prophecies is nowhere clearer than in predictive algorithms because they are the most opaque of self-fulfilling prophecies.

ALGORITHMIC SELF-FULFILLING PROPHECIES

Algorithmic predictions are scoring people for the purposes of loans, insurance, jobs, and more. Each of these is a self-fulfilling prophecy. As we apply algorithmic prophecies to more spheres, they shape more of our lives.

David Hume wrote that if God is omniscient and omnipotent, then it's hard not to blame him for evil. In the case of AI, if predictive analytics are partly creating the reality they purport to predict, then they are partly (and in some cases wholly) responsible for the negative trends we are experiencing in the digital age, including increasing inequality, polarization, racism, sexism, misinformation, and harm to children and teenagers from social media.

Mortgage Approvals

Crystal Marie and Eskias McDaniels fell in love with a $375,000 four-bedroom house in Charlotte, North Carolina. It had a neighborhood pool and a playground for their son. They had saved much more than they needed for the down payment. They had good credit scores—803 and 725—and each earned six figures, she in marketing and he at a pharmaceutical company. The monthly mortgage payment was less than what they had paid for rent for years. There was only one problem. Their mortgage was denied by an algorithm. The couple had spent $6,000 in fees and deposits—none of it refundable.

According to an investigation by the newsroom *The Markup*, lenders in the United States in 2019 were more likely to deny home loans to non-white people than to white people with similar financial characteristics on paper. They were 40 percent more likely to reject Latino applicants, 50 percent more likely to turn down Asians, 70 percent more likely to deny Native Americans, and 80 percent more likely to reject Black applicants than similar white candidates. Even high-earning Black applicants with *less* debt than whites were rejected more often than their white counterparts.[30] Crystal Marie and Eskias McDaniels are Black.

Why are algorithms so unfair? It may be because payday loan sellers tend to set up shop in neighborhoods inhabited mainly by people who are not white and where banks are less common. When people use these services, they end up with incomplete credit histories because payday loan services report only missed payments (not successful ones). Another reason is that most of these algorithms are fed with historical data, and racism is baked into it, as scholars like Safiya Noble, Joy Buolamwini, and Ruha Benjamin have shown.[31] Predictive algorithms identify correlations, and not being white is associated with having fewer opportunities.

Loan decisions are made by proprietary software designed by agencies like Freddie Mac and Fannie Mae. No one knows the details of how Fannie's and Freddie's software scores applicants. Their algorithms' decisions are not publicly available, and not even loan officers have any understanding or say regarding them. There are laws against discrimination, but the evidence suggests we need better enforcement.

When you are denied a loan, that lack of opportunity will further financially disadvantage you. As if that weren't bad enough, most financial institutions use the same or similar algorithms, so if you're rejected by one bank, you're likely to be rejected elsewhere. These systems are creating the economic conditions that make their initial predictions more likely. A person denied credit due to an algorithmic assessment will find it harder to build credit history, thus "proving" the original assessment correct, creating a vicious cycle. Prophecies become retrospective alibis.

Without investigative journalism that can compare different applications, it's impossible for an applicant to prove that they have been rejected for unfair reasons. Since there are no transparent criteria for getting a loan approved, the applicant is left powerless in a Kafka-esque system that depends on opaque probabilistic correlations.

A very similar situation happens with algorithms designed to select job candidates. A person scored as unemployable for obscure probabilistic reasons will find it harder to find a job, which will make them even more unemployable. That person not getting any jobs will count as "evidence" that the algorithm is accurate, but that is misleading, because the cause of that person not getting jobs could be the algorithm itself.

We want predictive algorithms to be accurate, but not through creating the reality they purport to predict.

Social Media Posts

We are attracted to the bizarre, the outrageous, and the violent, and social media algorithms are designed to maximize engagement. Posts that look like past popular posts get enhanced by the algorithm, which increases their visibility, making popular posts even more popular. Unfortunately, the content that hooks people the most tends to be toxic.

An AI that detects that a person stares for a microsecond longer at conspiracy theory posts will recommend more extreme content, potentially radicalizing that user's perspective through its own predictive mechanism. Someone who was merely curious can become a convert after being exposed to so much outlandish content that it comes to feel normal, thereby fulfilling the algorithm's prophecy.

Social media can drive vulnerable populations to serious harm and anxiety, as Jonathan Haidt has argued in *The Anxious Generation*. An algorithm that detects that a teenage girl clicks on posts related to diet can begin suggesting content that incentivizes eating disorders. Molly Russell was a fourteen-year-old girl from North London who died by suicide in 2017. Her father discovered that she had been viewing content related to self-harm on Instagram. Molly had been seeing "the bleakest of worlds," as her father put it.[32] Self-fulfilling prophecies can be a matter of life and death.

Predictive Policing

Predictive policing uses data analytics and algorithms to forecast criminal activity and allocate police resources efficiently. When police increase surveillance in neighborhoods flagged by algorithms, they detect more crime in those areas. This additional data then reinforces the algorithm's original prediction, regardless of whether there were indeed higher crime rates than in other, less policed districts.

Some of these algorithms were originally designed to predict earthquakes. It doesn't take a genius to figure out that there are many questionable assumptions behind the idea that a statistical model that works for earthquakes can be accurate predicting crime. Earthquakes tend to happen in some locations and not others because there are areas in the world where tectonic plates meet; when they move, they create an earthquake. Crime doesn't work that way.

Increased police presence and surveillance in certain neighborhoods can create tension with residents and erode community trust, leading to more confrontational interactions resulting in arrests, creating yet another feedback loop.

Predictive policing systems are trained on historical crime data, which reflects societal biases and discriminatory practices. If certain neighborhoods are historically over-policed, the algorithm will continue directing resources there, perpetuating patterns of disproportionate enforcement.

Among the surprises that statistics unveiled in nineteenth-century Paris was the sensational realization that the higher the educational level of a district, the higher its crime rate.[33] It's unclear whether these numbers were tracking a real effect or spurious correlations. Maybe those numbers reflect the fact that the rich tend to have more valuable things that are worth stealing. Or maybe the privileged did and still commit more crimes than the underprivileged. Perhaps those first numbers on crime and social class are the most accurate we have, because with statistics being in its infancy then, the highly educated were still not as adept at hiding in numbers.

Anecdotally, I have been jolted out of my naivete by the level of crime perpetrated within the privileged walls of elite universities, from academics (allegedly, of course) fudging their data, to stealing ancient papyri and selling them on the black market,[34] to sexual offenses of the worst kinds.[35]

As an undergraduate, when I was introduced to famous professors, I immediately felt a degree of trust in them by virtue of their social recognition. Many years later, the more prominent the professor, the more reserved my approach. If the personality in question is especially charming and self-confident, and presents himself as the most cited scholar in a field in which data is easily fudged, add another dose of caution. It's not part of my job description, but I have at times felt obligated to watch over my students like a hawk.

I suspect crime in high places is a combination of two effects. First, people who are willing to do *anything* to climb to the top are more likely to get to the top. That's why the incidence of psychopaths is greater among the top echelons of society, whether it's academia, business, or government. Second, those who get to the top acquire a sense of entitlement that makes them more likely to be the kinds of people you wouldn't want introduced to your daughter. The more they behave badly and learn that people will turn a blind eye, the more they trust they'll always get away with it, and the worse they behave.

While the crimes of the privileged go mostly unrecorded, crimes in poor neighborhoods are more likely to be logged and pursued. Disadvantaged and stigmatized populations are surveilled more aggressively,[36] they also have fewer resources to defend themselves, no one makes a fuss when they go to jail, and there is no powerful institution behind them worrying more about their own brand than the suffering of victims.

In the data economy, however, we are all being surveilled, even if the effects of that surveillance are distributed unequally. Our data is feeding the machines deciding our future.

SURVEILLANCE AND PREDICTION

Ancient soothsayers did not have the data economy at their disposal for espionage. Although the functions of predictions today are unsettlingly similar to what they were in the ancient world, the methods have turned much more invasive.

Your every search, keystroke, email, physical movement, and inter-action with other human beings is being recorded and analyzed by hundreds of shady companies that are inferring the most sensitive information imaginable, predicting your behavior, and trying to influence it for profit. You might have paid for your phone, laptop, and car, but they are not working for you. They work for the data brokers and tech companies they are sending your data to.

Data from various sources including public records, online track-ing, loyalty programs, app usage, purchase histories, and social media activity is used to create detailed individual profiles that can be used for targeted advertising, risk assessment, or personalized propaganda.

Data brokers all over the world have a data file on you. It con-tains information like your name, age, gender, income, education, and occupation. Data brokers, who behave with eerie similarity to vultures, track family life events like marriages, divorces, funerals, and pregnancies. They know who your family and friends are. They infer your sexual preferences, political tendencies, and health status, all without your permission. They track your location, which allows them to know where you live, where you work, how well you drive, whether you are buying drugs, or going to a psychologist, or attending a family planning clinic, whether you are having an affair (through patterns like two phones coming together in a hotel every week), and much more than you can imagine. They then "add value" by predict-ing what you'll do next and selling that to the highest bidder.

Privacy Is Power, which I wrote in 2020, remains painfully current. Perhaps the most significant development is generative AI, but for all its novelty, from the point of view of privacy, it's just another version of more of the same at scale. Large language models have ingested all the data on the internet available to them, including very personal data found in social media, forums, and databases of various kinds, and they are able to infer very sensitive data from less sensitive data.

Machine learning algorithms are good at sifting through moun-tains of data more quickly than ever and making profiles. That makes surveillance much easier. You might have the impression that what's happening when you interact with a chatbot is that you're getting information from it, but it is getting more information from you than you might suspect.[37] It's not only that you might be uploading contracts or other kinds of texts. It's also able to profile you by what you talk about, the time at which you connect with it, how you phrase sentences, and more. Language can be very local and idiosyncratic, and how you use it gives away information about your background.

Arguably, in contravention of legislation like Europe's General Data Protection Regulation and California's Consumer Privacy Act, companies like OpenAI cannot tell citizens what data has been collected about them from third-party sites for the training of their models, because they chose not to keep a record of the data they've used. Their data collection has been so indiscriminate that they don't know what data they have accessed. They cannot allow people to retrieve the entirety of their own data, because it is not held in accessible form. And, most important, they are unable to delete that personal data if a consumer requests it; the data is baked into the model, so OpenAI would have to throw away the cake and retrain the model without the personal data of that person, which would be too costly.

And yet on both sides of the Atlantic we are allowing companies and individuals to use these systems as if they were perfectly legal. The ugly truth that no one dares to say out loud is that we've allowed the prophets to break our rules, thereby setting their own new ones. The law is adapting to technology, instead of technology adapting to the law. Laws encode democratic agreements based on what kind of society we want to live in; unregulated technology doesn't. That an academic with no ties to big tech is writing these words is no coincidence, but there are fewer and fewer of us left; I'll get back to this problem (and solutions) in chapter seven.

I've researched privacy for more than a decade now, and every time I engage with the evidence of how much of our data is being collected and how it's being used, I get the same feeling I got when Forrest told me his wife had been told by a psychic that she'd die young. *How dare they.*

How dare these companies start tracking people from the time they are fetuses. What a sense of entitlement, to feel they have the right to infer sensitive information about our bodies and minds that we haven't given to them. And how dare they use that data to make predictions about whether we will live or die. How dare they make forecasts that will impact our life chances when those predictions are financially motivated and far from truths. How presumptuous of them, to treat us as if we were nothing more than pawns on their chessboard. How dare they treat us as things and not persons. Once they have our data, they don't even ask for our input anymore, as if our data could speak for us.[38] *How dare they.*

DIGITIZATION AND SURVEILLANCE

Prediction depends on surveillance, and according to the advocates of Bayesian machine learning, the more data, the better (although not really, as we will see in the next chapter). The new prophets think they have a problem: They are running out of data.

One way to get access to more data is to turn more of the analog world into digital form. That's why big tech companies continually push us to buy more things online, read in digital formats, scan the world with the cameras on our phones, try virtual reality, spend more time online. For all the success of the digital, most of our everyday life experience has not been translated into ones and zeros, the building blocks of the digital.

Companies say that technology is neutral, that whether it ends up having a positive effect depends on how we use it. That is a very convenient narrative, because it shields them from all responsibility and puts it squarely on our shoulders. Technology, however, is never neutral.[39] Someone designed it to do something. An artifact embodies values because the act of making it reflects a belief in the value of what it can achieve.

In philosophical jargon, artifacts have *affordances:* what an object designed by a person invites you to do. A chair invites you to sit on it. Of course, you can also use it as a projectile, but it wasn't designed for that. Affordances are the thread that connects the designer of an object with its user. Surveillance tools afford control; they invite people to keep a close watch on something or someone, and since human beings are highly social creatures, most of the time it's someone.

Digital tools are not neutral; they are built to surveil. To digitize is to surveil. I call that the *law of digitization* because there is currently no digitization without surveillance. Unless we radically reinvent the digital, everything analog that is turned into digital potentially enters the surveillance machine. The very term "data collection" is deceptive:

Data gets created, not collected. And the creation of data is a morally significant act because it changes the world. By turning things into data, we create a record. By digitizing, we make things trackable, searchable, and taggable: fodder for surveillance and prediction.

Tech companies' thirst for data is what makes them peddle the story that sharing is caring, that public exposure, which used to be considered impolite, is a virtue. Tell us what you feel, where you go, what you eat, what you think about other people, what worries you, what you're angry about. And if you don't want to tell? Then you must be hiding something. Big tech shames us into exposure for its own profit.

Digital records make society even more legible to all machines, whether it's a state, a corporation, or an AI, in the terminology of James C. Scott, the American political scientist.[40] Without a quantified record of human affairs, states are blind. When we turn the analog into digital, there is no more need to physically retrieve paper records; all digital records are in principle just a few clicks away. Digital records also make it easier for computers to process information: to identify people, catalog them, track their income and location, and predict their lives.

Many products on our devices (such as online documents and maps) were designed precisely with the creation of personal data in mind, to make our lives more trackable by machines. That they are useful to us is just the carrot to make us inadvertently cough up the data that is used to control us.

SURVEILLANCE AND DOMINATION

The act of surveillance is an act of despotism. It is subjecting someone to your gaze, inducing self-conscious emotions in them, pressuring them into conformity. For all primates, directly gazing at someone is a way of exercising power over them. Surveillance is a form of domination. We may be far away from the savannas we evolved in, but the stare of another still turns us into potential prey.[41]

Surveillance creates asymmetries of power, because the surveilled is at the mercy of the watcher. The more information the watcher has on their subject, the more power they have over them, and the easier it becomes to predict and influence their behavior.[42]

There is no surveillance for the sake of it. Surveillance makes no sense without prediction and an intention to control. Neither institutions nor predators surveil for fun. They surveil you to better guess how you'll act next and either influence your behavior or better respond to it.

If privacy is power, surveillance disempowers the citizenry. Remember Larry Ellison, the chairman of Oracle, the company that provides sensitive databases and predictive software? He has "predicted" a modern surveillance state in which "citizens will be on their best behavior, because we're constantly recording and reporting everything that is going on."[43] If there is one prediction you choose to defy, let it be that one.

LEARNING FROM OTHERS: THE SOVIET UNION AND CHINA

The Soviet Union

The Soviet Union was enthusiastic about planning; it involved making a grand prophecy, drawing up a blueprint for getting there, and following the plan. To achieve the ambitious goals required extensive quantification and surveillance.

Soviet leaders believed they could rationally design economic production by setting specific targets for every sector, from agriculture to manufacturing, through a series of five-year plans that began in 1928 under Joseph Stalin.

State planners would determine how many tractors, shoes, steel ingots, and other goods would be produced, allocating resources and setting production quotas for every factory and collective farm across the enormous Soviet territory. While this approach aimed to eliminate market inefficiencies and ensure equitable distribution, it resulted in catastrophic economic distortions.

The most notorious example of planning failure was agricultural collectivization. Moscow planners, most of whom had never plowed a field, consistently miscalculated agricultural needs. But speaking against the prophecies of the state could land you in the gulag, or dead, so everyone just went along with it, fabricating numbers when the reality and the expectations didn't add up, until bad news couldn't be hidden any longer.

Stalin's plan to consolidate small farms into large state-controlled collective farms led to massive grain shortages, resistance from peasants, and ultimately the devastating Soviet famine of 1932, killing millions.[44] So much for planning.

China

Contemporary China is the most extreme case of mass surveillance and state planning that we have ever seen. It has the ambition of previous authoritarian regimes paired with the capabilities of digital technology.

Around 700 million cameras watched the Chinese population in 2024, about one camera for every two citizens.[45] Police officers roam the streets in smart sunglasses with a built-in camera linked to Sky Net that can identify people. Sky Net is the name of China's surveillance system, as well as the name of the AI that controls the world in the dystopian future of the *Terminator* movies. Why use a euphemism when you don't have to?

If you happen to be part of a minority, you're automatically treated as a suspect, if not a criminal. The government can install a camera in the corner of your living room; it records sound, too. Police officers store your DNA in a government database after swabbing your mouth and drawing a blood sample. Your voice and face are likewise logged by recognition software. If you're a woman, the government can forcefully sterilize you, or demand that you take a birth control pill every day at noon. A minder might be appointed to you, a government official who will live with you, sometimes in your own bed, if your husband has been mysteriously disappeared by the state. If he makes advances on you, resisting him could expose you to disappearing too.[46]

Rules are enforced at random. You can be hauled away to a concentration camp for the most subtle of "infractions," like having dinner with someone with a low social ranking.[47] The camps function outside the criminal system. Since no one detained there is formally charged with a crime, there is no knowing for how long they'll be captive. Since 2017, an estimated 1.8 million people have been interned in these camps for harboring "terrorist thoughts." It is the largest internment of minorities since the Holocaust.[48]

The government has placed devices on people's doors that allow the police to record surveys. To enter a grocery store or a gas station, you have to scan your ID card at the entrance. Only if the computer deems you a "trustworthy" citizen are you allowed entrance.

Even if you're not a minority, life is hard under a surveillance state. The internet is heavily censored. A Chinese citizen doesn't have access to learning about the Tiananmen Square massacre in 1989. Wikipedia is banned. Its replacement, Baidu Baike, claims to be "open and free" but doesn't have an entry for 1989, a whole year obliterated.[49] All calls are scanned in real time with voice recognition; if the system spots a "criminal," it alerts the police.[50]

WeChat, China's most popular messaging app, is closely watched by authorities. An alliance between tech companies and the government makes sure that every message is filtered and clocked, and no criticism of the government is tolerated. Citizens also use WeChat to book trains and flights, make appointments with doctors, call taxis, pay for their water and electricity bills, hire a house cleaner, match with dates, get fast credit, and manage investments—all valuable data to surveil and predict. No one uses cash anymore.

The objective of such close surveillance is control of the population, and prediction is a crucial piece of the puzzle. Professor Zhang Zheng in Beijing compares China's social credit system to Western data brokers and credit scores like those of Experian, but with even more data. It includes how you handle your finances, but also how you treat your parents, and your relationship to the Chinese Communist Party. We shouldn't be satisfied with surveilling people. "Looking at their future is more important," he says.[51]

If you are blacklisted by the party because the system has predicted trouble, you have no future. Your friends and family cannot call you (if they try, they'll get a message saying that you've been placed on the list of "trust breakers"). You cannot get a loan, or buy a train ticket, or stay at a hotel. You are not visible on a dating app. To keep your social points up, you have to avoid playing too many video games, changing addresses too often, or being friends with people with a low score. Watching videos of speeches of the party leader earns you points.

Companies are the closest allies to the Communist Party, with no choice but to abide by a law that coerces them to give up whatever data the governments asks for.[52] Tencent developed WeChat. Megvii (backed by Sinovation Ventures, a venture capital firm set up by Kai-fu Lee, a Taiwanese businessman and writer) and SenseTime provide image recognition. Hikvision specializes in surveillance cameras and is partly owned by a government agency. Huawei (roughly meaning "China shows promise") was started by a former military engineer and has been associated with the surveillance of minorities.[53]

Western companies are complicit too. In the early '00s, Yahoo, Microsoft, and Google signed a pledge agreeing to censor any information the government deems inappropriate.[54] Apple banned an app used by Hong Kong protesters to track police movements; it also limited file sharing on iPhones in China after protesters used AirDrop to spread leaflets critical of the party; it didn't make available a privacy feature designed to obscure a user's web browsing from internet service providers and advertisers; and it ceded control of its Guiyang data center to the government.[55] The American semiconductor companies Intel and Nvidia provide the chips necessary to process Chinese surveillance data at the Urumqi Computing Center.[56] The DNA sampling equipment to track citizens came from Thermo Fisher Scientific, the American medical technology company.[57]

China is selling surveillance around the world. I've encountered Hikvision cameras in streets in the U.K., Europe, and Mexico. Belgrade has struck a "safe city" partnership with Huawei to install eight thousand cameras around the city. Scores of cities in some 150 countries are plastering their streets, airports, schools, and buildings with Chinese cameras.[58] Having Chinese spyware as the infrastructure of cities not only carries the risk of empowering illiberal regimes; it also gives Beijing both access to sensitive data and the power to easily shut down a city's operations. That's not a smart idea, however much we wish to call these cities "smart."[59]

EXPLOSIVE INGREDIENTS

If you're lucky enough to live in a liberal democracy in times of peace, you might think that an authoritarian regime is so distant that it might seem impossible. You'd be wrong. Democracy is fragile. Do you think that the rise of digital technology, with all the surveillance and prediction involved, is a step toward greater or lesser freedom? If your government did something egregious tomorrow, would you be afraid if you or one of your children took to the streets to peacefully protest? Are you afraid of looking up certain politically sensitive (but perfectly legal) topics online?

In *Seeing Like a State,* James C. Scott identifies four key elements involved in the most tragic episodes of state-initiated social engineering:

> The legibility of a society provides the capacity for large-scale social engineering, high-modernist ideology provides the desire, the authoritarian state provides the determination to act on that desire, and an incapacitated civil society provides the leveled social terrain on which to build.[60]

The first element may be the most important one, because without it there is no grand-scale forecasting: Legibility involves standardization, quantification, digitization, and the "administrative ordering of nature and society." The second element involves having high confidence in technological progress, to the point of it becoming an ideology, an unquestioned faith, a cult. The third element is an authoritarian state, but I don't see why it couldn't apply to authoritarian corporations as well, or to alliances between despotic governments and corporations. Finally, a weakened citizenry that doesn't rebel against authority is the last needed ingredient for tragedy to ensue.

Digital technologies have cracked civil society. We spend our days looking at our phones instead of at each other. Our screens spy on us and manipulate our perceptions. Tech companies have lied to us time and again, eroding our trust in them, in our government, and in our fellow citizens. Digital technologies make it harder to defy authority: We are losing the possibility of anonymous protest, of sharing a reality unmediated by personalized digital content, as well as social and political skills. The more we introduce identifying technologies into the private and public spheres, the higher the fear of having to pay a price for dissenting. In China, to be uncooperative is to sacrifice your life.

The four ingredients for disaster are there, like kerosene, matches, gunpowder, and fire sitting side by side.

DESPOTIC CORPORATIONS

You might think that because companies are doing most of the surveillance and prediction, there is less to worry about. Governments are scarier than companies. It is only the government that has the monopoly of violence in modern states, as Max Weber put it. It's governments that can put you in jail.

I am sorry to be the bearer of bad news, but it doesn't make much sense to distinguish between corporate and government surveillance. Whatever data is collected by companies can end up, and often ends up, on the screens of government officials, and whatever data is collected by governments ends up being analyzed by corporations.

Further bad news is that companies can be as threatening to justice and freedom as the worst of governments. At its peak, the East India Company was the largest corporation in the world. It had twice as many soldiers as the British government. Among its many sins were slave trafficking, facilitating the opium trade, exacerbating rural poverty and famine, and looting India.[61] A senior official of the old Mughal regime in Bengal wrote in his diaries, "Indians were tortured to disclose their treasure; cities, towns and villages ransacked; jaghires and provinces purloined." Companies can act like vicious states. As the historian William Dalrymple puts it,

> We still talk about the British conquering India, but that phrase disguises a more sinister reality. It was not the British government that seized India at the end of the 18th century, but a dangerously unregulated private company headquartered in one small office, five windows wide, in London, and managed in India by an unstable sociopath—Clive [the first British governor of Bengal].[62]

In the eighteenth century, if you were a huge company wanting to grow, you conquered other lands with boots on the ground. In the twenty-first century, you conquer the analog world by translating it and transforming it into a digital format. Amazon conquers a slice of the world when it persuades you to read a Kindle book rather than a paper one. When companies like Amazon, Google, Microsoft, and Palantir are getting military contracts with governments, it's not a big stretch to compare them to the East India Company. People who used to work for intelligence agencies are all over the data economy.

The market dominance of big tech is astonishing. No other industry in history, and no other government, had been able to influence so many lives as any one of these behemoths, let alone all of them together. Amazon, Apple, Nvidia, Microsoft, and Meta are worth trillions each, making them wealthier than most countries around the world. These companies' market capitalization exceeds the capitalization of the whole of the U.K.

What never ceases to astonish me is the extent to which companies and governments place their trust in big tech. Policymakers and CEOs around the world act starstruck around the new prophets, and go about obeying their prophecies like soldiers responding to commands from above in the hierarchy. Even more astonishingly, governments and companies everywhere I look depend on big tech to run their infrastructure. They put their most sensitive data on the servers of the prophets. Start-ups use Google Docs to plan their future. Don't they know Google has access to that data and is in the business of making start-ups go away?

Law firms and doctors and governments use Microsoft Teams to share documents and talk about sensitive information. Even though Microsoft denies that they use this data to train AI, haven't people read Microsoft's vague privacy policy and realized that there is nothing there that specifically bans them from doing that? If I were an alien and had just landed on Earth, I'd infer from the faith placed in big tech that these were the most trustworthy companies in history. But I'm not an alien newly arrived; I've read the news; I have a collection of their scandals, tagged by type.

Too many people opt for a combination of cynicism and gullibility, readily arguing that the world is burning down and that other people are hell, while at the same time trusting the most questionable types with our democracies and our lives. But both cynicism and gullibility are complacent responses to a historical moment that is asking us to rise to the occasion. Isn't a combination of idealism with street smarts more appropriate?

ON THE ACCURACY OF PREDICTIONS

Computer scientists and engineers around the world are working hard to build more accurate prophecies. When it comes to predictions about people, that is a mistake. Accurate forecasts about individuals are not good news; they are not a scientific advancement, but a symptom of tyranny.

Hannah Arendt warns us that "domination does not allow for free initiative in any field of life, for any activity that is not entirely predictable." She also points out that tyrants tend to announce their nefarious intentions in the form of infallible prophecies. Instead of owning up to his intended actions of making war and assassinating Jewish people, Hitler announced to the German Reichstag in January 1939, "I want today once again to make a prophecy: In case the Jewish financiers . . . succeed once more in hurling the peoples into a world war, the result will be . . . the annihilation of the Jewish race in Europe."[63]

The future is unwritten. When predictive accuracy about persons increases, it is not because we are discovering the future but because we are determining it. If we're able to accurately forecast people's fates, we're that much closer to turning human beings into things. That's too high a price to pay for accurate predictions.

We'd do well to heed Arendt's warning: The "method of infallible prediction, more than any other totalitarian propaganda device, betrays its ultimate goal of world conquest, since only in a world completely under his control could the totalitarian ruler . . . make true all his prophecies."[64]

In the previous chapter, we analyzed the first pitfall of prediction: confusing truth with likeness to truth. This chapter dealt with the tendency of predictions to bring about the reality they prophesy. The following chapter is about how, no matter how hard you surveil and predict, not everything is predictable, and the more you think everything is predictable, the greater at risk you are of being crushed by the unforeseeable.

THE CRYSTAL BALL IS CRACKED

*Why the unforeseeable is never going away,
and how forecasts can increase risk*

Jeanne Calment, born in 1875 in Arles, France, remembered working at her father's shop when she was fourteen years old and selling colored pencils and canvases to Van Gogh. He was "ugly like a louse," she said.[1]

Jeanne and her husband, Fernand, were double second cousins; their paternal grandfathers were brothers, and their grandmothers, sisters. They had a daughter, Yvonne, in 1898, who died from tuberculosis in her thirties. Yvonne left behind a husband, Joseph, and a seven-year-old son, Freddy. In 1942, the Calments went to visit some friends in their country house. Fernand gorged on cherries during the visit; Jeanne had a few. The cherries were tainted with chemicals, and Fernand died from poisoning.

After her husband's death, Jeanne shared an apartment with her son-in-law, Joseph. Some twenty years later, in 1963, Joseph died too, and that same year her grandson, Freddy, was killed in a car accident. Jeanne had lost all her immediate family. She was 88 years old.

When Madame Calment was 90, and with no heirs left, her notary, André-François Raffray, offered her a deal. She would get a monthly revenue of 2,500 francs (about $500 at the time) until her death. In return, the notary would inherit her apartment in Rue Gambetta when she died. Raffray must have thought he had landed a steal. The old woman could die any day, and he'd get a very well-located apartment almost for free. But Madame Calment had different plans.

When Jeanne was 100, she was still spryly riding her bicycle around town. "I had to wait 110 years to become famous. I intend to enjoy it as long as possible," she said.[2] Jeanne Calment outlived her notary too. Raffray had paid his monthly checks for thirty long years, only to die on Christmas 1995, at the age of 77, having paid Calment more than twice the value of an apartment he never got to live in. As Calment put it dryly, "In life, one sometimes makes bad deals."

On the night of Raffray's death, Calment was dining on foie gras, duck, and cheese.[3] She boasted of eating more than two pounds of chocolate per week. (When I read her story in the press, decades ago, I sent it to my mom to justify my chocolate consumption; she wasn't persuaded.) Calment kept receiving her monthly check from Raffray's widow, until she finally bowed out, at the ripe age of 122 and a half—the oldest person ever whose age has been verified. Well played, Madame Calment, well played.

Twenty years after Madame Calment died, Valery Novoselov, a Russian geriatrician and former doctor in the army, said he thought she had been a fraud; he claimed that she looked too young for her age in her photos. He teamed up with Nikolay Zak, a mathematician who used a database of centenarians and concluded that the probability of someone living to age 122 was "infinitesimally small." A theory ensued: What if Jeanne was really Yvonne? What if there had been a switch between mother and daughter? Could Madame Calment have been a con artist?

When Lauren Collins, from *The New Yorker*, traveled to Arles to get to the bottom of the story, she realized that for Calment to have been a fraud, she would've had to co-opt the whole city, including the notary who ended up paying her twice the value of her flat. Raffray had administered the marriage contracts of both Jeanne and Yvonne; there is no way he could've confused them. And Jeanne had enemies in Arles—she begrudged Communists—who would've been all too happy to turn her in had she tried a scam.[4] So it turns out a fraud is even more unlikely than a woman who ate two pounds of chocolate a week living to be 122.

Life is unpredictable, in wondrous and tragic ways alike, in the mundane and the extraordinary. People make plans and the gods laugh.

Think about the best and the worst things that have happened to you. They are likely also the most unpredictable in your life. Think of all the near misses you've had, the times you've almost died, or how you were almost not born, or how you almost didn't get to meet one of the most important people in your life. This chapter is about why the most significant events of life are unforeseeable.

There are at least five sources of unpredictability, or predictive troubles.

First, there are *data troubles,* also known as imperfect knowledge. We are plagued by bad data and limited data. And data will always be limited, no matter how much of it we collect. By collecting as much data as possible, we can get closer to authoritarianism, but not perfect predictability.

Second, there are *social troubles.* Social worlds are composed of many interconnected moving parts, with reflexive patterns, and subject to power laws. We'll discuss the book industry as an example of fat-tail distributions that refuse to yield to prediction.

Third, there are *scientific troubles.* As the philosopher Karl Popper argued, for strictly logical reasons, it is impossible to predict the course of history because it is impossible to predict future scientific discoveries and technological innovations that will greatly influence history.

Fourth, there are *coincidental troubles.* Fortuna, the Roman goddess of luck, is everywhere.

Fifth, there are *ironical troubles.* Our very predictions are making the world more unpredictable.

We'll take these in turn, and end with a reflection on how the ultrarich are preparing for the future and what that tells us about their prophetic enterprises.

FIRST: DATA TROUBLES

When Jeanne Calment was born, the life expectancy for a French woman was forty-five years. Based on the data available, there was no way that Raffray could have guessed his longtime acquaintance would happen to end up having the longest recorded lifespan.

The doubts about Calment's authenticity only make the example better. Every single day, we make risk decisions based on questionable information.

Bad Data

About 70 percent of data is made up. That's a bad joke; a half joke, really. The much less funny empirical finding is that most research studies cannot be replicated; if you redo the experiment, you don't get the same results.

The journal *Nature* conducted a survey in 2016 and found that out of 1,576 respondents, 70 percent of scientists had tried and failed to reproduce another researcher's experiment, and more than half had failed to reproduce their own experiments.[5] Replication efforts in psychology have found that only about half of studies are reproducible.[6] Paul Glasziou, then chair of the International Society for Evidence-Based Health Care, and Iain Chalmers, one of the founders of Cochrane (an organization that conducts systematic reviews of medical research), estimated that 85 percent of all health research is wasted by not being reproducible, resulting in $170 billion in losses worldwide per year.[7]

In 2005, John Ioannidis, a professor at the Stanford School of Medicine, published a paper titled "Why Most Published Research Findings Are False."[8] One problem is the use of $p < 0.05$ as a sign of "statistical significance." In plain English, what that means is that scientists have chosen an arbitrary cutoff point below which they assume that their results are a product of a real effect they are observing, and not a fluke, but the cutoff point is way too high. A $p < 0.05$ result is one you can expect to see due to chance alone one in twenty times. The lower the p value, the more likely it is that the effect is tracking something meaningful.

As it stands, all you have to do to find some interesting result is to do twenty studies, and you are almost guaranteed to see something, even if you are not tracking a real effect. That is, there will be a coincidence roughly in one study out of twenty that looks like an interesting finding but is merely a fluke. In one study, researchers "proved" that listening to a Beatles song made people younger. The paper, by Joseph Simmons, Leif Nelson, and Uri Simonsohn, is called "False-Positive Psychology," and it showed how statistics can take a coincidence—that the participants in the group listening to the Beatles happened to be younger than the group listening to another song—and make it look like a causal relationship that is not there.[9]

They showed that if you torture the data long enough, it will confess to *anything*. The worrisome part is that it is a common practice to fish for an apparent result through the magic of statistics. "Everyone knew it was wrong, but they thought it was wrong the way it's wrong to jaywalk," said the researchers. But "it was wrong the way it's wrong to rob a bank."[10] When everyone is doing something wrong, it gets normalized. But that doesn't make it one least bit right; just look at Nazi Germany for an extreme example.

Another dirty statistical trick is to stop collecting data if your p value drops. That is, if having more data weakens the effect, then choose where to stop collecting data. Yet another problem with scientific studies is that negative findings are rarely published. More appallingly, the vast majority of scientists don't share their data.[11] Even the researchers who state in their articles that they are willing to share their data often don't when asked.[12] Sometimes there are good reasons not to show the data—like privacy—but most times there aren't.

In 2023, an article was published in *Nature Human Behaviour* showing that replicability could rise to 86 percent when following scientific best practices, which include transparency of the data and preregistering hypotheses, measures, and analyses (registering what you will do before you do it). Less than a year later, that same article was retracted.[13] The authors hadn't followed best practices.

Here's another one: Dan Ariely, from Duke University, and Francesca Gino, from the Harvard Business School, were star researchers in behavioral science. They made their fame researching dishonesty. Can you guess where this is going? Joseph Simmons, Leif Nelson, and Uri Simonsohn uncovered that their data had been tampered with.[14]

Just imagine that you were someone hungry for fame and with no moral scruples. You observe these practices as an undergraduate in behavioral economics, psychology, or business. You know that exciting research will get you published in top journals, and that if you don't publish, you won't get a job in academia. You know that if you repeat a study twenty times, you are bound, if only by coincidence, to find some exciting correlation. You know that you'll never have to give up the data, so maybe you save yourself the trouble of collecting it and just go ahead and fabricate it, or you collect data and tweak it to fulfill your needs. Even if, by some extraordinary reason, your data was discovered, if you know what you're doing, it's not hard to make the numbers "compatible" with random sampling so that even the data detectives could not easily uncover the fraud. The risk is so low, and the payoff so large.

You start with a spectacular (made-up) result, and sure enough you get published in a top journal. Soon, a superstar in your field cites you, and the social contagion starts. Perhaps the superstar is your colleague or a friend who wants to do you or your mentor a favor. All it takes is to be on the reference list of a paper authored by the right person and other people will cite your work too, until it becomes accepted convention that your work is a must-cite in any paper about the topic. Often the most cited work on a particular topic is the worst work in that field, but it is authored by someone with power and connections, or simply someone who got there first.

The more famous you become, the safer you are. Eventually, someone asks you for your data. You know you don't have to show it to them, but it would look better if you had a good excuse not to. Maybe you claim it is proprietary. Maybe you set up your own consulting gig, and what you offer is data analysis using your secret sham database. If you get tenure at a prestigious university, and advise Fortune 500 companies and governments, no one will challenge you because too many important people have an interest in your not being outed as a fraud; everyone would be embarrassed. It's much easier to continue the ruse. Don't fall for slick con artists with suggestive graphs in fields in which it's easy to manipulate the data.

There are much more innocent reasons for why we are swamped with bad data: from measurement errors to measuring the wrong thing, making wrong assumptions, conditions changing, and much more.

Academia is one of the most important bastions of democracy, and I'll defend it in the next chapter. That's precisely why it's important to clean it up. The truth is that a concerning proportion of research studies should not be trusted, and that much of science depends on people trusting and citing their colleagues and friends without fully knowing whether what they cite is true. It's not only that it's possible to publish potentially false results in science; it's that *science's incentives reward fraud more than truth.*

The good news is that we know, roughly, how to fix the problem. I won't get into it, but there are some useful references in the notes if you want to follow that trail.[15]

The upshot is this: We are making predictions in medicine, policymaking, and business on the basis of data that has a high probability of being false. That increases the chances of making bad predictions and being blindsided by what you didn't see coming. Any prediction based on scientific studies that have not been replicated risks being seriously wrong. An optimist might retort, Let's just fix the data! Let's. There's value in that. But data can be misleading even when accurate.

Misleading Data

Data is always somewhat misleading in the same way that all models are wrong but some are useful. Whether they are scientific theories, statistical analyses, or physical representations, all models simplify reality; if they didn't, they wouldn't be models. In the process of simplification, features of reality are lost. Models abstract away complexity and nuance in favor of workability. A model is valuable when it is useful, not when it is true. If a model allows you to better visualize a space, or to communicate an idea, that's good enough. Data, like a model, is a map of territories. We tend to forget to distinguish between maps and the real world.

If you build a model of an airplane, it will be smaller than the actual plane. No matter how precise you make the model, that means that its aerodynamic properties are simply not going to be the same as the real airplane, for reasons having to do with size and weight. If you made the model the same size as the airplane, then it wouldn't be a model; you'd be making the airplane itself.

Say you are studying how people feel about privacy. If you collect data on people's behavior, it mostly seems as if they don't care about privacy. Most people say yes to cookies (the little bits of software that collect your data online), and regularly give up personal data to companies. But that is misleading.

When you ask people whether they care about privacy, most of them say "yes." The behavioral data doesn't capture the reasons why people click "yes" to cookies, which are likely a combination of it being too inconvenient to say "no" to cookie banners (the option to say "no" is usually more cumbersome and less visible than the option to say "yes"), feeling busy and overwhelmed, thinking that their data is out there already, and not trusting that companies won't track them anyway.

When Apple gave people the choice to say "no" to data collection on their iPhones, about 96 percent of Americans chose to protect their privacy.[16] It was a convenient onetime click, and it felt like an actual choice. If you had predicted that people would say "yes" because they had said "yes" to cookies in the past, you would've been wrong.

Say you want to study gamblers, and you stumble upon a curious piece of data: The trope about beginner's luck is empirically true! Explanations flood your mind: Maybe it's helpful not to know too much about games, or maybe it's the overconfidence that someone who has never lost brings to the table. You might even be encouraged to become a gambler for a while, until the beginner's luck runs out. That would be a costly prediction. The data is a product of *survivors' bias*. The people who are lucky at the beginning of their gambling careers are more likely to stick with it, whereas those who were unlucky are likely to choose a less painful occupation. The data of the dropouts doesn't get collected, because they are not at the gambling table anymore.[17]

The same effect of survivors' bias can be seen with mutual fund performances. When you see numbers about how well these mutual funds are doing, it's easy to forget that the statistics often exclude failed funds, which artificially inflates average returns. That's why financial platforms include a disclaimer that past performances are not indicative of future results; the fund you choose might become part of the silent cemetery of dead funds. Caveat emptor.

—

Suppose you wanted to study crime and you find a peculiar piece of information: Cities that sell more ice cream have higher rates of crime. If you're tempted to write an email to the mayor of your city recommending to ban ice cream immediately, don't. Hot weather drives both effects. That's called a *confounding variable*: Whenever there is a correlation between two phenomena, there might be a third factor driving both.

Here's another example: You can "predict" the rate of Nobel laureates in a country by how much chocolate it consumes.[18] Brazil has fewer than 0.1 Nobel Prize winners per 10 million people; the United States has about 10 laureates per 10 million people; the U.K. has almost twice as many; Switzerland has more than 30 per 10 million people.

The correlation is so strong that you'd expect it to occur only once in ten thousand cases by chance, if there were no true effect. But that's the magic of statistics; in a world that increasingly collects more data, we are also increasingly running into more spurious correlations. Recommending that all countries binge on Swiss chocolate will not produce more Nobel laureates.[19]

I could fill a whole book with these kinds of examples, but you get the gist: Data never tells the whole story, just as the map never captures the whole of the territory, and there is always more than one way to interpret data. The eternal optimist might think that this problem could be solved if only we got more data.

When Less Is More, or Why More Data Isn't Always Better

Those who seek to understand the world through data naturally tend toward wanting more of it. The dream is to have all data. The theory is that we get things wrong in science because we work with samples, but a sample of the population might not be representative or might leave out outliers from which we could learn. If you could study the *whole* population, you wouldn't have to worry about questions of generalization. The idea of big data has old roots, from Laplace's demon and Bayesians' view of using all data available, to the philosopher Rudolf Carnap's principle of "total evidence."[20]

The first thing to note is that if we try to acquire as much data as possible about the population, we're well on our way to creating a police state through mass surveillance. But let's set aside that minor issue for the time being. Let's also set aside for now how the more data we create—that is, the more records we produce—the more computing power we need to process it, and the more we contribute to climate change. Even if you try to create data on everything, you will fail, because data is the map, and maps cannot have the richness of the territory; they wouldn't be useful if they did.

The second thing to note is that not all data is relevant. Even if we manage to somehow get rid of inaccurate and false data, more data is not better if the extra data tells you nothing about what you want or need to know. Getting reams of data about law-abiding citizens will not help authorities find criminals.[21] Adding more hay to the haystack just makes the needle harder to find. Much of the data we are creating and are exposed to is just noise. As Nate Silver puts it, "The noise is increasing faster than the signal."[22] The signal is truth, while noise is the data that distracts us from the truth.

Third, and relatedly, there are some cases in which the more data we have, the worse our predictions. That's why the philosopher of science Karl Popper argued that we shouldn't go around trying to confirm our theories, because we can always find data to fit them (also known as cherry-picking). Rather, we should try to find fault with our theories.[23] Even then we might fail; sometimes the data is simply not there (for example, because the historical event that will disprove our theory hasn't happened yet). Bayesian thinking (and machine learning) require a world in which the past is a reliable indicator of the future so that data can provide dependable priors. But uncertainty has a will of its own and doesn't always bow to the Bayesian imperative.

Heuristics

In 2008, Google announced that it had found a quicker way to predict the flu than the traditional method of the Centers for Disease Control, which relied on counting the number of flu-related doctor visits in the United States and took a week or two to compile the information. When we fall ill, we tend to look up our symptoms on a search engine. Google engineers analyzed around 50 million search terms, then tested 450 million different models, and designed an algorithm that used 45 search terms. Google called its prediction "nowcast," because it had a lag of only one day.

But when the swine flu broke out in the spring of 2009, Google Flu Trends missed it. It had learned from previous data that flu infections happened in the winter. Google's engineers went back to work, and they bet on adding complexity to the algorithm. They increased the search terms from 45 to about 160.

Once more, the algorithm seemed to be incredible at prediction at first—until it wasn't. It greatly overestimated the impact of the flu between 2011 and 2013. The model was tripping up because some people searched for symptoms just out of curiosity, not because they were sick, and Google couldn't tell one motivation from another. After seven years of tinkering with the algorithm without success, Google Flu Trends shut down in 2015.

Gerd Gigerenzer and his team had a hypothesis: that one single data point could be more useful than amassing as many data points as possible. They were right. They came up with a heuristic for predicting the flu that beat Google's algorithm: predict that this week's proportion of flu-related doctor visits will equal those of one week ago.[24]

Heuristics are knowledge in distilled form: shortcuts, rules of thumb. They are easy and fast to apply, and animals excel at them—especially human beings. A heuristic is a simple rule that ignores part of the information available to make inferences that can be more accurate than complex analyses. In situations of uncertainty, simple heuristics can beat mountains of data.

The term "heuristic" comes from the ancient Greek word *heuriskein,* meaning "to find." The word is also related to "eureka!" famously attributed to Archimedes as an exclamation of discovery. As Vitruvius tells the story, the king of Syracuse gave Archimedes a problem to solve: to find out whether the craftsman who made the crown he had commissioned had used pure gold or an alloy. The king had given the goldsmith a lump of gold to make the crown, but rumor had it that the goldsmith had stolen part of the gold and replaced it with a silver alloy.

Archimedes was lowering himself into a bath when he had an epiphany: He understood the principle of buoyant force. Gold is denser than silver, which makes a piece of gold of a certain weight smaller than a piece of silver of identical weight. If the goldsmith had used silver instead of gold, then the total volume of the crown would be greater than the original gold. You can compare the volumes of objects by observing how much water they displace.

Archimedes had figured out that he could take a piece of silver, and another of gold, both weighing exactly the same as the crown, submerge them in water, observe how much water overflowed in each case, and compare them with how much water overflowed with the king's crown. He ran naked through the streets of Syracuse shouting "Eureka!" The king had been conned; the crown was not pure gold, because it displaced more water than the same weight in gold.

Don't think of heuristics as second-best options that we resort to only because of our cognitive limitations. If the alternative is a "long slog," and is no more accurate, the shortcut is the better option.[25] When a baseball player catches a ball, she doesn't perform a set of differential equations. Rather, she fixes her gaze on the ball, starts running, and adjusts the running speed and trajectory so that the ball remains squarely in her gaze.[26] That's a heuristic, and it's much better than the alternative.

Heuristics are all the more valuable in contexts in which creating data has costs—in time, attention, computing power, natural resources, or other goods we value like privacy—but can be preferable even in (mostly theoretical) contexts in which information is free.[27]

Heuristics are especially helpful in situations of great uncertainty. Economists at the Bank of England devised simple decision trees that outperformed complex models for calculating banks' capital requirements.[28] The economist Joseph Stiglitz and his colleagues found that fast and frugal heuristics work better than more data-heavy alternatives for making predictions about economic environments that are highly complex.[29]

Heuristics have also been shown to work better in managerial decisions—the kinds of strategic choices that are the bread and butter of any company. For instance, research suggests that using heuristics produces better outcomes in decisions about whom to hire than trying to add up and weigh all available information.[30] Simple approaches are also likely to have further benefits by being easier to understand and communicate. In turn, greater clarity can contribute to greater accountability and fairness. Moreover, reducing data collection increases privacy.

When It's Not About Data

Finally, there are some things in life that are not about data. Recently I stumbled upon an example while reading the news. "How Better Data Could Lead to Better Sex," read the article title.[31] I laughed out loud; the barista in my favorite coffee shop sent me a curious look.

After reading the article, I found the idea slightly more plausible: If we knew more about what others' experiences are like in the bedroom (or the kitchen), we'd have a better sense of what is possible, and we'd be less inhibited. That sounds reasonable. The article ended by lamenting the little funding there is for such research.

What the article missed is that for us to have that data, a change in culture would be necessary. And it's the change in culture, not the data, that could lead to better sex. Some of the hypotheses behind why we don't have more data are that sex tends to be mired in shame, and that there are interests in covering up just how much violence is perpetrated in sex.

Those problems don't get solved with more data. If we manage to change the culture into one in which talking about sex with our loved ones comes naturally, sexual violence is shunned, and body positivity encouraged, we wouldn't need the data after all. I doubt that the most proficient lovers are spending their free time poring over the latest scientific articles on sex.

Good sex involves trusting and being trustworthy, being good at communicating with another person, being capable of intimacy and making yourself vulnerable, being comfortable in your own skin, being good at reading and respecting body signals, mastering the joy of consent,[32] being able to be in your body and not in your head, and many other sensuous wonders like scent and touch that have very little to do with data—at least with data in the form of squiggles on a spreadsheet.

Conversely, more data could lead to worse outcomes. If most people in your community practice a sexual act that you don't find attractive, getting confronted with that fact might make you feel pressured to do it because everyone else is doing it. Teenagers are feeling pressured to submit to practices that have been normalized through porn.[33] Or let's say you have a sexual quirk that no one else is doing; it might lead you to feel stigmatized and stop you from asking your partner whether they'd like to try it.

Perhaps novels are the best kind of sex education you can get. A survey can tell you that most people kiss, but not what the kiss feels like—whether it's tentative or untethered, too wet or just right, too fast or too slow. It says nothing about whether the kiss is an act of lust, domination, or love. Spreadsheets are silent on whether fingers are slender, whether hands are warm, or whether the drummings of your heartbeats are so close to each other that you can no longer tell which belongs to whom.

Veiled under the protection of fiction, authors will tell you what sex is like, what good sex tastes of, the joys of its textures and smells, and what bad sex feels like. Part of you is right there in the novel, close enough to gain experience and understanding—to taste hunger in someone's breath, to notice the room getting damp and warm, to feel your skin turn to goose bumps—yet distant enough to be safe. And it doesn't even count as cheating.

Novels tell you about the aftermath of sex, about the glowing giddiness of the day after good sex, and the soul-crushing shock and panic left in the wake of sexual violence. Novels reveal what defies the ignorant boundaries of what seems logical to us—how sometimes you can know someone in one touch, or how we can act as if nothing has happened when the worst has.

And novels let you be; there is no bell curve approving your normalcy or identifying you as an outlier. You can take the experience of another human being and do with it what you will. If you're reading a paper book, no one is watching or judging (e-books, on the other hand, are spying on you). I recommend starting with D. H. Lawrence's *Lady Chatterley's Lover* and ending with Sally Rooney's novels. We don't need more data. We need more great novels and great readers, as well as decent human beings, and a cultural milieu that fosters intimacy and respect.

More data is not necessarily better. Data pushes us away from being our very particular selves. Maybe the best sex is the one that finds the exact sweet spot in which very peculiar people in very particular times of life in a specific mood that day can mutually enjoy themselves.

It's not only sex. There are crucial aspects to being a good human being—a good friend, a good parent, a good boss—that no amount of data will magically improve and for which true empathy, goodwill, sensitivity, courage, humility, kindness, curiosity, and other virtues are necessary. *The world is driven by values, never by data.*

Why There's No Such Thing as Data-Driven

Chris Anderson, then editor of *Wired,* wrote in 2008 that the sheer volume of data we were creating would render theory obsolete and the scientific method dispensable.[34] "Correlation supersedes causation," and "with enough data, the numbers speak for themselves," he wrote, and swaths of philosophers of science turned in their graves or their office chairs.

Anderson is an eloquent representative of mottoes that are often heard in podcasts, consulting companies, and, of course, tech companies. He is also wrong. There is no such thing as data-driven. Data is dead; it's the narratives, explanations, and relations that we impose on the data that make it come alive.

Even more fundamentally, as the philosopher Karl Popper pointed out, before we can collect the data, we need to have some kind of direction; otherwise we wouldn't be interested in collecting that particular kind of data; the *problem* always comes before the data.[35]

Data are artifacts. Anderson writes about data as if they were natural kinds, which exist independently of human beings; we didn't invent natural kinds like clouds, corals, or comets. But data are not natural kinds; data were not things in the world before the advent of human beings. We don't collect data, even if we talk that way; we *create* data. Data are designed by people, built to serve our curiosity and interests. Of course, data correspond to something in the world. Data can track natural kinds, just like a scale can track your weight, but that doesn't make the scale the driver of anything. Every data point is like a small mirror that tells us something about the world, but we build the mirror, and we position it one way or another. *We* drive the data through our values, through what we want to know or achieve, through our will to power. And then we claim that the data is the driver to cover our tracks.

A paper by the cognitive scientist Abeba Birhane and colleagues analyzed one hundred highly cited machine learning papers and identified fifty-nine values that motivated and were uplifted by AI research.[36] When a consultant or a policymaker tells you that their approach is data-driven, consider hiring someone who can identify the underlying values. While philosophers tend to be mostly useless at much of life's tasks, we are nonetheless trained to identify bad arguments, to the infinite annoyance of those around us—especially those who stand to make money off faulty reasoning.

The data never tells the full story. Data can be informative or misleading, or a bit of both, but they don't explain anything, and they don't make decisions. Even in the case of decision-making AIs, ultimately the buck stops with the people designing and implementing them.

When someone says they are taking a data-driven approach, one of two things might be going on. They might know they are being led by their own agenda and taking you for a ride. They can always find, fabricate, or torture the data to justify their goal. Or they might *think* they are being driven by data. What they are actually doing is being led by unconscious theories and biases, or they are following someone else's agenda.

When playing around with chatbots, I'm amused by how they are willing to recognize problems with tech such as bias, but tend to finish their speeches pointing out how accurate, sophisticated (a word they overuse), and complex the technology is, and how all the problems are either fixed or getting fixed. If I buy their narrative, I'm peddling the interests of OpenAI or Anthropic or Google. When we follow data unreflectively, we are often fulfilling tech's agenda.

The bottom line is that you need good judgment, above and beyond data, to make good decisions. And if you don't own your decisions and take control of the helm, someone else will gladly take the ship in the direction that's convenient to them, along with your money and freedom.

When no one was thinking about the world in terms of data, collecting data could provide some competitive advantage. But "data-driven" approaches have become the norm in the valley of the blind. The one-eyed person is the one who can still use her own judgment.

Data can be false, misleading, or incomplete, which may lead us astray when making predictions. That's true of events that follow a normal distribution. The bell curve is so impressive that we tend to interpret all phenomena in those terms.[37] But not all phenomena fall within the curve. And if we assume we are tracking a normal distribution kind of event when we are not, we can get things catastrophically wrong.

Fat Tails

The height of human beings follows a normal distribution. Two-thirds of people will cluster around the average, and outliers don't stray too far away from it. The shortest living person in the world is an Indian woman called Jyoti Amge who measures 24.7 inches (62.7 centimeters); the tallest is a Turkish man called Sultan Kösen who measures 8 feet, 2.8 inches (251 centimeters). That's a huge variation, but it's limited. Kösen had to receive lifesaving surgery to stop him from growing. The biological facts of our bodies can tolerate only so much variation. If you become too tall, your heart will not have the strength to pump oxygen to your brain.

If I tell you that there are two people whose heights add up to fourteen feet, what do you think is the most likely breakdown of their heights? They are both probably around seven feet tall. People who are taller than eight feet are rare enough that it's safer to bet for the average.

But not all phenomena follow the statistical distribution named after Gauss. Earthquakes, terrorist attacks, extinction phenomena, wealth distribution, and book sales, among others, follow a fat-tail distribution. The normal curve looks like a bell, large in the middle with its edges (or tails) thinning at the margins, tapering toward zero. Fat-tail distributions are called that because they have thick and elongated tails. Take the example of the size of a city: While on one extreme the population is zero, the upper limit is unclear; in this case, the fat tail is one-sided.

While we know roughly what the limits of human height are, we don't know what the limits are with fat tails. There might be an upper limit to how severe an earthquake is, but we don't know what it is, and because it's so extreme, it might as well be infinite for practical purposes.

The most extreme fat-tail examples are called *power law distributions*. Wealth distribution follows a power law. The poorest person in the world owns $0 (they might also be in debt, but let's bracket that for simplification). The richest person has hundreds of billions of dollars. The tallest and shortest people in the world can take a photo together; to put the poorest and the richest people on the same graph, you have to use a logarithmic scale, which compresses data by displaying values on a multiplicative rather than additive scale.

The concept of standard deviation becomes meaningless in power law distributions because there are no typical variations. Averages and medians in normal distributions are informative; not so for power law distributions, because neither measure captures the distribution's defining characteristic: its extreme inequality.

In other words, our statistical normal models break down in the face of power law distributions. The incentives are toward ignoring extreme outliers because they make our models ugly, because they reveal the limits of our predictive abilities, and they remind us just how risky and unfair life can be.[38]

SECOND: SOCIAL TROUBLES

Suppose you take a random sample of two Americans who jointly earn $10 million per year. If wealth followed a normal distribution, you could expect them to earn about $5 million each. As things stand, it's much more likely that one of them earns around $9,950,000 and the other one around $50,000.[39] Book sales—as well as music sales, art in general, and other social phenomena—follow a similar pattern.

Books, Revisited

Suppose I tell you that two books by two different authors have jointly sold 1 million copies. What is the most likely breakdown between them? Something like 5,000 and 995,000.[40] The world of publishing is as unequal as they come. A minuscule number of books account for most sales. Even though a publishing house can maybe manufacture a bestseller through a self-fulfilling prophecy, it also depends on spontaneous bestsellers to make a good profit. The latter tend to be more profitable than the former because they've cost the publisher much less.

Aware of the unpredictability they face, publishing houses have asked external consultants how to increase their profit. The result is a running joke in the industry:

> Millions of dollars have been spent, untold numbers of hours have been chewed up analysing and arranging and rearranging. And eventually a report . . . gets dropped on the CEO's desk and there is a cover sheet on top. And in some form or another within the first four bullet points, what it announces is that the management consultants involved in this study have finally solved the riddle of publishing. They've worked it out and here it is. You've given us 5 million dollars, here's the answer: only publish bestsellers.[41]

Publishers know that they often lose money on the books for which they have given advances of anywhere between $350,000 and $750,000. "Consultants will recommend that we only buy books at the ends of that spectrum. The advice assumes that we know what a book is worth, but we don't. We're making a bet given our best guess, but there are few safe bets," says one New York editor.

Some unexpected bestsellers will come about as a result of word of mouth, like J. K. Rowling's Harry Potter series. If you love a book and want it to become part of the public discourse, gift it to your friends, recommend it to your book club, and talk about it. It might also come about thanks to a gutsy editor and publisher who are willing to invest more on a book that they believe in than the gods of data would prescribe, because while the downside is limited and under control (the amount invested in marketing), the potential upside is unlimited.

In addition to bestsellers, publishers profit from a robust backlist—old books that sell consistently well. Unlike the capricious front list (the new books), you can count on the backlist, its income being steady and predictable. Some publishers earn more than half of their revenue from backlist books.[42] Marcus Aurelius, William Shakespeare, and Jane Austen stand a great chance of selling well next year.

But if it's hard to predict which book will be a bestseller, it's even harder to predict which will become a backlist title. Data analysis is not going to help you tease out the extraordinary contemporary book that will become a classic. The bestsellers of today are often like the popular kids in high school who end up not doing much with their lives, while the relatively unsuccessful books can end up like the inconspicuous teenager who goes on to change the landscape in their field as an adult.

The philosopher Arthur Schopenhauer had a tense relationship with his mother, Johanna. She was a successful writer, and the first German woman to publish without a pseudonym. She authored novels and travel journals. Her second book, about a family tour of Europe, was a hit, and soon she became the most famous woman in the country. She held a renowned salon attended by Goethe and the brothers Grimm, among other celebrities.

When Schopenhauer finished his doctoral dissertation, he sent it to his mother's publisher, F. A. Brockhaus. The publishing firm saw nothing of value in the book, but it wanted to keep Johanna, one of its best-selling authors, pleased. Except Johanna was not pleased. She found the manuscript incomprehensible and told her son that it was unlikely to ever sell a copy. Furious, Schopenhauer responded that his work would still be read long after her "trashy novels" were forgotten.[43]

He was right. Good philosophy books as well as novels that tackle existential dilemmas stand a better chance to survive the test of time because the most fundamental questions of life never go away, and gaining clarity on them makes life more livable. The dilemmas of ancient and modern people are similar.

The editors at Brockhaus got things thoroughly wrong: They didn't please Johanna, and they made the best publishing decision of their careers, even though it seemed like the worst. It took about three decades for Schopenhauer's books to be appreciated, but he was lucky enough to see it. Nietzsche's works, on the other hand, were not widely read while he was alive. Hume famously said of his *Treatise* that it fell "deadborn from the press." And Franz Kafka couldn't have imagined the impact his works would have.

The next time you meet a writer whose books are not being read widely, be like Phoebe from *Friends* and tell them, "Oh, my God, you're not even appreciated in your own time. I'd give anything to not be appreciated in my own time!" You'd be doing something very similar to most prophets: taking data, committing fallacies (affirming the consequent, in this case), and providing anxiety relief.

(What would you prefer, if you had to choose: to have success in your lifetime and have your work be devoured by oblivion after your death, or to be a failure in your lifetime but have your work be wildly successful and have an impact for centuries after you're gone? I think I'd choose the latter, but please don't let that stop you from recommending this book.)

We'll return to the backlist in chapter nine.

THIRD: SCIENTIFIC TROUBLES

The most important scientific breakthroughs are the least predictable ones. Alexander Fleming discovered penicillin (the first antibiotic) by accident when one of the samples of bacteria in his laboratory got contaminated with a fungus while he was on holiday. The thinker who wrote about this argument in the clearest and most logical way was Popper.

Sir Karl Popper was perhaps the greatest philosopher of science of the twentieth century. He was born at the beginning of the turn of the century in Vienna, when the city was the center of the intellectual world. Figures like Sigmund Freud, Gustav Klimt, Arnold Schoenberg, Moritz Schlick, Rudolf Carnap, and Ludwig Wittgenstein gathered at coffeehouses such as Café Central and Café Landtmann to discuss art, psychology, and philosophy.[44]

Popper trained first as a cabinetmaker, then as a teacher in mathematics and physics; then he did a PhD in psychology, and finally ended up becoming a giant of the philosophy of science. In 1937, he took a position in New Zealand to get away from the fraught political situation in Europe. While the Nazis annexed Austria, he wrote a critique of totalitarianism—his war effort. In 1949 he became professor of logic and scientific method at the University of London.

Popper argued that "for strictly logical reasons, it is impossible for us to predict the future course of history," and no scientific or rational method can overcome this obstacle.[45] History is greatly influenced by scientific and technical knowledge. From the invention of the wheel to the discovery of how to make fire and beyond, scientific knowledge has changed how we live and die. Knowledge has even changed our biology. Making fire helped us eat cooked food, which changed our bodies.

We cannot predict through any rational or scientific methods the future of our scientific knowledge for the very simple reason that if we could predict it, we would already know it. To predict future events, we need to predict technological innovations, but if we could envisage the artifacts of the future, we would have already invented them. We cannot know today what we will know tomorrow, and therefore we cannot predict what will come the day after tomorrow.

Technology is advancing at an exponential rate in the twenty-first century, argues Azeem Azhar in his book, *Exponential*. The faster pace of change, the drastic transformations brought about by globalization, the combination of new technologies, and opportunities to learn from an increase in products make it all the more impossible to predict the future (even the not-too-distant future).[46]

—

The limit of science is that it cannot predict itself. "No scientific predictor—whether a human scientist or a calculating machine—can possibly predict, by scientific methods, its own future results," wrote Popper.[47] If you are predicting the future of humanity, then you're not doing science. Whatever you think you are doing is closer to reading animals' entrails. Most likely, you're playing power games.

Popper calls those efforts "scientistic," borrowing the term from F. A. von Hayek. Scientistic methods imitate the methods and language of science without being scientific.[48] They treat social groups as if they behaved like physical or biological bodies.[49] Hannah Arendt wrote that "the scientificality of totalitarian propaganda is characterized by its almost exclusive insistence on scientific prophecy," with spokespeople pretending "that they have discovered the hidden forces that will bring them good fortune in the chain of fatality."[50]

History responds to singular events, not to laws or generalizations. And the human factor is the most uncertain of all. Every attempt at predicting human behavior is also an attempt to control it, which in its extreme leads to authoritarianism. Tyranny, argues Popper, amounts to the "omnipotence of the human factor"—the whims of a few people prevailing, temporarily.[51]

FOURTH: COINCIDENTAL TROUBLES

Fortuna was the Roman goddess of luck, made immortal by Boethius, the last of the Roman philosophers. Boethius was also a senator, a consul, and a *magister officiorum* (one of the most senior administrative officials in the Roman Empire). From the very highest echelons of power, Boethius fell to the very bottom. He became disliked for denouncing corruption in the government and was imprisoned and condemned to death.

Boethius wrote his most known work, *The Consolation of Philosophy,* while awaiting execution in prison. He argued that Fortuna is capricious and no one is immune to her whims; only virtue can give you the kind of happiness that is not endangered by the unavoidable vicissitudes of chance. The book went on to be one of the most influential of the Middle Ages.

Fortuna is whimsical; it takes away and gives for no apparent reason. What if Boethius hadn't discovered corruption in the senate? What if he had had a cold and missed the meeting of the Royal Council in Verona where he felt obligated to defend someone who was being falsely accused of a crime, which prompted his arrest?

As family legend has it, my grandmother, in her journey fleeing from fascist Spain, was scheduled to take a flight from New York to Mexico City. My family had saved money for the trip to spare her a train ride because she was eight months pregnant. My grandmother had already found her seat on the airplane, but a flight attendant, alarmed at the size of her belly, asked her to get off the plane and get permission from the airport's doctor to fly. The airplane left without her; it ended up crashing, killing everyone on board.[52]

It's not only our individual lives that are subject to the goddess of fortune; flukes can often be found behind major political and social events.

Archduke Franz Ferdinand was first in line to the Austro-Hungarian throne. In a visit to Sarajevo, a member of Young Bosnia, a Serbian nationalist movement, threw a bomb at the archduke's car, but it bounced off the car and missed its target. There were a few people injured, but the archduke and his wife were unharmed. What if they had called it a day then? They didn't.

The archduke decided to attend a scheduled meeting with local officials. After the meeting, instead of accepting the offer of a tour of the city, Ferdinand decided to visit the people injured by the bomb. His driver, not knowing the way, took a wrong turn on the way to the hospital. The Austrian military governor of Bosnia yelled at the driver that he'd taken a wrong turn. The driver stopped the car, and when he began to reverse, the engine stalled and the gears locked.

By chance, a co-conspirator of the first attacker was standing in front of a delicatessen precisely on that street. As the archduke's car came to a halt, he took the opportunity and shot Ferdinand and his wife at point-blank range. Their deaths are widely recognized as the spark that ignited World War I. All because of a wrong turn.[53]

"Cleopatra's nose: had it been shorter, the whole aspect of the world would have been altered," wrote Blaise Pascal. If Cleopatra's nose had been shorter, Mark Antony wouldn't have been awestruck; had he not been awestruck, he wouldn't have sought an alliance with Egypt against Octavian; had he not made that alliance, the Battle of Actium would not have happened, Octavian wouldn't have consolidated his power over Rome, and our world today would be different.

If Napoleon hadn't had a cold (or a severe case of hemorrhoids, according to other versions) during the Battle of Waterloo, he wouldn't have delegated command to Ney, who had suffered the shooting of four horses from under him, which led him to make mistakes, like sending the Garde Impériale two hours too late.[54]

And on and on, history unfolds, greatly influenced by the goddess of fortune, who introduces flukes that forever alter the path ahead. The more we realize how vulnerable we are to chance, our desire to know what the future holds intensifies, making prophets more popular, and encouraging us to rely on predictions.

FIFTH: IRONICAL TROUBLES

The irony is that an overuse of prediction makes the world more unpredictable.

Why Predictions Create a False Sense of Safety

Predictions can create a false sense of safety that leads us to take more risks. In 87 BCE, the consul Octavius craved for more power. He overthrew his colleague Cinna, with whom he disagreed on most issues, and drove him out of the city. But Cinna wasn't going to go away quietly. He took control of the Roman army stationed at Nola, and persuaded the exiled general Marius to help him fight Octavius. Battle ensued. According to Plutarch, the astrological prophecy that had assured Octavius of his safety and success was found on his dead body.[55] Would Octavius have been more careful without the soothing prophecy?

Cicero, in his attack against astrology, reminds us that Pompey,

Crassus, and Julius Caesar all thought that they would die at home of old age; prophets had told them so. They all suffered violent and untimely deaths instead. Perhaps they might've been more cautious if they had been less confident of their fates.

It is because we want to be reassured that we are easy to manipulate, and prediction industries tend to thrive, irrespective of their helpfulness or lack thereof. In business and government, data and AI have become alternatives to critical thinking and creativity, as well as ways to dodge accountability. If a report suggests a course of action purportedly supported by numbers and graphs, it's easier to follow suit than to question the conclusion and take responsibility for the risk.

Forecasting, in business, policymaking, and beyond, is used to relieve anxiety and to make money much more than to make good decisions. The main promise of prediction for prophets is power, but the main product they sell is anxiety relief.

There are no experts in the future. Plato's Socrates makes this point in *Laches*. There are statisticians and historians, but only a time traveler could know what the future holds. Some people are better than others at forecasting, just as some people have better judgment than others; the political scientist Philip Tetlock calls them "super-forecasters," but even they are very far from knowing what awaits us tomorrow.

If you read the press, you'll notice, however, that experts are constantly making statements about what lies ahead. I wonder how many times they volunteer an opinion about the future and how many times it's journalists or the public who ask about it. I try to resist that question, reminding people that the future is unwritten. But because there is such a demand for predictions, experts who readily talk like prophets are more sought after, thereby feeding the forecasting frenzy.

When Tetlock set out to study expert predictions, his conclusion was concerning. The experts in his research had done barely better than random chance at foretelling political events. Worse, the more famous the expert, the worse their predictions tended to be.[56]

That means that not only are there no experts in the future; there are no experts who know about the future of whatever they are experts in. An expert in AI might know quite a bit about their field, but that doesn't give them a competitive advantage in knowing about what the future of AI is. A true expert is the opposite of a prophet.

We don't like uncertainty; that's why we ask experts to tell us what the future looks like. But we can't know the future, and denying the undeniable rarely works out well in the long run. Acknowledging that the most important events are unforeseeable is unpleasant, but ironically, denial robs us of the best that uncertainty can bring.

Selling Risk Management While Increasing Systemic Risk

If you run a business or work in government, you'll have noticed that many companies are trying to sell you predictive AI and that one of their main selling points, in addition to efficiency, is risk management. The new prophets tell us that we can use AI to predict who will default on a loan, which employees are thinking of leaving, or when our equipment will need maintenance. Machine learning can manage supplier risk by predicting which suppliers might be unreliable. Algorithms can purportedly predict trends, including what consumers will want to buy in the future. The new oracles say that they can predict pandemics, crime, social unrest. But, much like how surveillance sheds a light on its surroundings but not on itself, AI creates its own risks that it doesn't identify, much less manage. And those risks are much more serious than the ones it claims to minimize, as we will see.

Ulrich Beck's *Risk Society* was published in 1986 but is as relevant to the age of AI as any technology book published today. In modern society, Beck argues, the production of wealth is systematically accompanied by the production of risk.[57] Artificial intelligence produces risk. While wealth accumulates at the top of the social hierarchy, risks accumulate at the bottom.

In the absence of isolable clear causes, no one takes responsibility for the risk created. Technological disruptions are taken as a fact of life, as if they were our God-given fate, as opposed to artifacts designed and implemented by human beings. Citizens of democracies sleepwalk into the conjured-up designs of the small group of people who are creating AI. The fiction that a particular technology is predestined absolves creators, citizens, and policymakers of any responsibility. The more we use AI, the more we feel complicit in its disruption, and the less likely we are to demand better products.

Making victims feel complicit is one of the oldest pages of the tyrant's playbook.

The risks brought about by technology transform from social costs (negative externalities) into business opportunities. The new prophets can sell their technology as the necessary poison and the remedy. The commercial exploitation of risks follows a thin line between revealing some risks and concealing others, with the gravest risks being the unacknowledged ones.

Existential Risk

Not to mention existential risk when writing about the systemic risks caused by AI would feel like an oversight, but I write about it reluctantly. Existential risk—the risk that AI will lead to human extinction or near extinction—has already received more attention than it deserves, to the neglect of more realistic, immediate, and grave problems.

Technology executives entrench their position as prophets and accrue further power through focusing on predictions, and the further away from the present, the better, the more outlandish and unverifiable their claims can be. The new prophets prefer talking about existential risk as opposed to problems like mass surveillance and unjustified self-fulfilling prophecies because it benefits them. It puts trouble in the future, which relieves them of solving the problems of today. It distracts the public.

Existential risk strikes a chord because it plays with tropes from centuries of fiction that scare us. When we're scared, we make bad decisions. When we feel vulnerable, we tend to put ourselves in the hands of the people who offer us the illusion of safety. And the prophets are only all too happy to position themselves as our saviors (from their own inventions!).

Talking of existential risk makes AI sound more powerful than it is. It suggests that AI is or will be smarter and more capable than human beings in every way. Existential risk justifies all kinds of actions like tech companies keeping their algorithms secret. Some people have gone as far as recommending extreme surveillance, including "informant networks" and "frequent police raids" to curtail risks—as if creating a police state did not carry an existential risk! Authoritarian regimes can be much more dangerous and powerful than in the past thanks to the surveillance technologies we have developed.

Existential risk also turns AI into a myth, a purportedly dangerous autonomous agent, thereby deflecting attention from the real dangerous agents: the new prophets. The most powerful invention of tech companies is not technology but narratives. They are great storytellers. It's our job to question them and not fall for their tales.

In December 2024, Geoffrey Hinton, the Nobel Prize winner who is often referred to as the godfather of AI, said that there is a 10 to 20 percent chance of AI wiping out humanity in the next three decades.[58] Earlier that year, he came to Oxford to give the Romanes Lecture at the Sheldonian Theatre to warn about that very risk. I asked him a question.

"If you truly believe that AI is so dangerous, and you played a fundamental part in its design and implementation, if you were to do it all over again, knowing what you now know, what would you do different?" Hinton paused for a second. "I wouldn't do anything different. *Je ne regrette rien,* as the song goes. Recently there was a *New York Times* article which suggested I did, but it was the journalist pushing for that answer."

That struck me as a curious response. There are a few ways to interpret it, and I don't find any of them especially flattering. Is it that he doesn't genuinely believe AI is as dangerous as he claims? Or, if he is sincere, is it that he values humanity so little that he thinks we should all be put at risk of extinction for the benefit of developing a machine? Or is it simply that he's motivated by fame and profit? (In his answer, he did go on to mention how he still has Google stock, half jokingly.)

We don't need AI to face existential risks. Human beings are perfectly capable of generating them on our own. We have climate change and authoritarian regimes knocking at our doors. Tyrannies are more likely to engage in wars. Since the development of nuclear weapons, wars have included existential risk. Authoritarian regimes are also more likely to kill sectors of their population—an existential risk to those human beings who end up chewed by the jaws of the machine. How about we focus on those risks?

For now, AI is a flashy autocomplete. Another way to put it is that it is more likely that we seriously hurt each other *using* AI before we can meaningfully say that *it* hurt us. The greatest danger lies in the human beings who design and implement the AI, who shirk responsibility for their actions, who damage democracy, who are building a mass surveillance system that leads to an increase in social control and a loss of freedom, who pontificate about existential risks as if they were being prudent, and who increase risks while claiming to reduce them.

One of the ways in which the use of predictive analytics is increasing systemic risk is by pushing risk onto individuals. It's been happening for years, and no one is talking about it.

Pushing Risk onto Individuals

In December 2024, Brian Thompson, the CEO of UnitedHealthcare, was assassinated as he was walking toward the annual investor day. His company's health insurance division had earned $281 billion in revenues the previous year, and he had been compensated with $10.2 million for his job.

The internet flooded with stories of people whose premiums had shot up or whose claims had been denied, or both. "Thoughts and deductibles to the family." "Unfortunately my condolences are out-of-network."[59] Those were the mild comments on social media. A killer had become a hero. A human being had died, his family mourning, and a significant number of Americans were cheering. Meanwhile, the company that provides security services for 80 percent of Fortune 500 companies said its phones were "ringing off the hook."[60]

A year before the assassination, a lawsuit was filed against United-Healthcare alleging that the insurance provider used AI to deny coverage to elderly patients who were on a Medicare Advantage plan, despite knowing that the algorithm had a 90 percent error rate. The lawsuit argued that UnitedHealthcare continued to use the AI because it banked on people's "impaired conditions, lack of knowledge, and lack of resources to appeal the erroneous AI-powered decisions."[61]

Health insurance companies are usually opaque about what kinds of algorithms they are using and when they started using them, but health-care professionals started noticing more claim denials between 2019 and 2020. In 2023, patients sued Cigna, another health insurance company, over an algorithm that reportedly rejected more than 300,000 claims in two months.[62]

Part of what is going wrong with insurance is that it is gradually moving away from the pooling of risk to a pay-for-your-own-risk model.[63] In the past, insurance companies would group individuals whose medical costs were combined. Some of those individuals would be healthier than others, but premiums were homogeneous (or adjusted for age) under the assumption that pooling risks allows the higher costs of the less healthy to be offset by the lower costs of the healthy. It's a model of solidarity.

With the development of big data and artificial intelligence, insurance companies feel more confident in calculating individual risks. Depending on your genetic makeup and data from your smartwatch, smartphone, supermarket bill, and so on, an insurance company can calculate how risky *you* are and charge you accordingly. Under such a model, insurance is no longer about pooling risk. Rather, individuals are increasingly paying their own way, including for risks that they have no responsibility over (such as bad genes, or being born in a disadvantaged neighborhood).

But an individualized model betrays the function of insurance. Paying your own way means you no longer benefit from the law of large numbers—you'd be better off paying directly for whatever misfortune befalls you (instead of paying for that plus for insurers as traders). It's also unfair. The most disadvantaged members of society are the ones who carry the heaviest load of risk; charging them more compounds unfairness in a phenomenon known as the Matthew effect, named after Matthew 25:29 in the Bible: "For whoever has will be given more." Predictive algorithms are notorious for worsening Matthew effects.[64]

With an individualized model, insurance companies will earn more in the short term, getting to keep the cheapest customers and getting rid of the expensive ones. But that creates an insurance underclass that makes *everyone* less safe.[65] The chief executive of the financial watchdog in the U.K. has warned that AI risks making some people "uninsurable."[66]

We've seen this kind of effect before. The 2008 financial crisis was largely brought about by the same mechanism. Banks sold mortgages liberally, including to people who didn't stand a chance to pay them back, and then sold those risks to other institutions. In the end, the bulk of the risk was placed on the shoulders of individual mortgage payers who could not carry it. Collapse was a matter of time. We're making the same mistake, and it's not only in insurance.

In the worst corners of the gig economy, employees have to be "ready to work" at all times, without knowing when they will work, whether they will work, or how much they will earn, all the while not being paid for the time they are waiting for work to be allocated to them by an algorithm in an app. On top of irregular and last-minute scheduling, some shifts are too short (sometimes lasting for an hour or less) or too long (lasting for more than eight hours), too early in the morning or late at night, often with no breaks. What happened to stable shifts with a steady income? We can blame the increasing use of predictions.

Traditionally, the risks of fluctuating consumer demand were mostly borne by companies who had to pay employees in restaurants, factories, or warehouses even during slow periods. Accurate predictive analytics can tailor shifts to fit expected amounts of work, all at the expense of workers, who no longer have stable shifts or wages.

The greater use of predictive analytics comes partly from a desire to minimize risk from the perspective of the institutions using such tools. A warehouse doesn't have to spend money on staff if work is slow, thereby minimizing its risk. But often, risk does not get minimized overall; it gets pushed around, and the tendency is for it to land on the shoulders of the most vulnerable of individuals. It's weakening society's resilience, which ultimately depends on individuals thriving. The workers in warehouses are left without the safety of a steady income they can depend on, but they still need to pay rent every month to stay afloat.

We should protect the worst off among us, if not out of kindness and solidarity, at least for the safety of social stability, on which we all depend. Individuals have fewer resources to absorb the hardships of risk than communities.

Leaving No Leeway

Prediction pushes us to minimize redundancy. Institutions are always under pressure to be more efficient. Resources are scarce, margins are tight, competition is fierce. One tempting way to be more efficient is to predict exactly what you will need, and to secure just that, leaving no room for extras. For retail companies, one selling point of predictive analytics is the supposed ability to predict what your customers will order before they do so you can have in stock those products, and those products only.

During the COVID pandemic, the U.K. had fewer hospital beds than its neighbors. It might have seemed like a good idea while there was no public health emergency.

While avoiding waste is, all things considered, a good thing, there is danger in having no leeway. Redundancy is like buying insurance. It can seem wasteful when everything goes well, but it can be lifesaving when disaster strikes. That's why our bodies have fat reserves, and two kidneys and two eyes. Optimization is good for the bottom line in the short run, but it makes us brittle.

Building No Fire Doors

In the analog world, we build watertight compartments in ships and fire doors in buildings to contain disasters. In the digital world, we are connecting everything to facilitate data collection and AI. Your laptop is connected to your phone, and your phone is connected to your laundry machine, your "smart" lights, your car, and anything that we can plug a chip into. If tech enthusiasts get their way, they will make your brain the next point of connection. It's about the worst idea we've had.

Every new connection in a system is a potential point of entry for malicious actors. If all your devices are connected, it means that hackers can potentially gain access to your phone (a relatively secure device, if you have a good one) through your smart kettle (most likely an insecure system). If all our national systems are likewise connected, a cyberattack could bring them all down. We are expanding our attack surface and getting rid of fire doors.

In 2020, Russian hackers gained entry into software called SolarWinds that is used to manage information technology infrastructure. Hackers inserted malicious code into legitimate software updates for a product called Orion that is used by about thirty-three thousand public and private clients. Ironically, Orion was supposed to be a watchdog: It monitored the performance of computer networks through analyzing devices, traffic, and connectivity to detect problems, prevent outages, and identify security threats. But it became a Trojan horse. Because the SolarWinds software was deeply integrated into the networks of thousands of organizations, this single point of compromise allowed attackers to access and spy on an enormous web of connected systems. The impact was massive: Around eighteen thousand organizations downloaded the compromised updates, including multiple U.S. government departments, Microsoft, Intel, Nvidia, and Cisco.[67] Our reliance on interconnected systems can turn trusted relationships into security vulnerabilities.

Destroying Variety and Volatility

Variety, or diversity, serves as a fundamental safeguard against catastrophic failure in both natural ecosystems and business environments. In nature, we see this principle at work in crop genetics. The Irish potato famine of the 1840s devastated the population precisely because farmers had come to rely on a single variety of potato, the Irish Lumper, which was vulnerable to potato blight. In contrast, the indigenous farmers of the Andes maintained thousands of potato varieties, providing resilience against any single disease or pest.

This same principle applies in modern business ecosystems. The dominance of a few large cloud providers—AWS, Microsoft Azure, and Google Cloud—creates systemic risks. When AWS experienced outages in December 2021, June 2023, and October 2025, they affected countless businesses that relied on its services, including the

New York Metropolitan Transportation Authority and *The Boston Globe*.[68] Technological monocultures, while efficient, create widespread vulnerabilities.

As AI language models and recommendation systems become more prevalent, they risk creating a cultural monoculture. When multiple news organizations use similar AI tools to optimize headlines and content, it can lead to a convergence in narratives. Similarly, when businesses use AI recommendation systems trained on similar data, they might guide consumers toward increasingly uniform choices in books, movies, and other products.

Because AI relies on statistical analyses, it tends to predict as successful what will be smack in the middle of the normal curve, except when it tries to maximize engagement in social media, in which case what is "successful" is whatever is outrageous. In most contexts, the algorithm selects the well-trodden path.

Have you noticed how many coffee shops look the same, whether you're in Oxford, Madrid, or New York City? It's the tyranny of AI. Social media algorithms on platforms like Instagram, Yelp, and Google Maps reward a particular kind of aesthetic that photographs well and generates engagement. Success online generates success offline: Google Maps highlights a few cafés and prioritizes them in searches, and people end up going to them, instead of looking for the quainter options. Searching for a high-rated coffee shop on an app is quicker and more convenient. So what if the place has no soul?

Café owners and baristas compare their businesses with those that are trending on apps like Instagram and copy the style in the hope of achieving a similar popularity online that might fill their tables.[69] Algorithmic recommendations (for jobs and for design) tend to favor the average and the bland. If we adapt to the algorithm, it tends to flatten us out, erasing our individuality,[70] as well as the diversity from which spring a richness of ideas and potential solutions to problems.

With less variety, there is less short-term volatility, but more fragility. Consider a tropical rainforest versus a commercial tree plantation. The rainforest as a whole remains remarkably stable year after year, despite (or rather because of) containing thousands of species competing intensely and experiencing individual boom-bust cycles. The plantation, in contrast, appears more stable day-to-day but is much more vulnerable to threats like pests. This pattern appears in economic systems too. Think of restaurant turnover in New York City, where specific establishments frequently fail but the restaurant scene as a whole remains vibrant.

If AI drives standardization and reduces variety, we might see less day-to-day volatility but greater systemic risk—much like monocultures that appear stable until they fail catastrophically.

Risky Times

Far from minimizing risk overall, algorithmic prediction is *increasing* risk for society. In the age of big data and AI, "precise forecasts masquerade as accurate ones," as Nate Silver puts it.[71] Numbers give the illusion of accuracy because they are precise. Enthusiasts of AI offer detailed numbers for risks that they are comfortable discussing while ignoring ones that could be ruinous and for which we have no reliable numbers. We tend to ignore the risks that we cannot measure, even when they pose the greatest danger to our ways of life. Numerical predictions incentivize risk-taking because they create the illusion that all risks are manageable.[72]

While the products of wealth are tangible, the worst risks lie dormant. The imperative of economic growth prevails because the visible wins over the invisible.[73] Paradoxically, risks succeed precisely because we fail to see them.

It seems that only the new prophets are aware of the risks they are introducing for all of us. They are counting on being able to wait out the storm in their luxury bunkers (an oxymoron if ever there was one) if things go wrong.

THE BUNKERS OF THE ULTRARICH

When the writer Douglas Rushkoff accepted an invitation to give a talk in the middle of the desert to a group that described itself as "ultra-wealthy stakeholders," he wasn't sure what to expect. Two billionaires and three multimillionaires wanted to talk; no speech required. They had questions. "Which region will be less affected by climate change? Which is the greater threat: global warming or biological warfare? How long should one plan to be able to survive with no outside help? Should a shelter have its own air supply? What is the likelihood of groundwater contamination?"

These people were building or buying their own bunkers to survive the event. What event? Oh, you know, the environmental disaster, social unrest, nuclear explosion, deadly pandemic, or technological mayhem that will make the world unlivable.

Some bunkers for sale have pools, with light that imitates natural sunshine. In addition to thinking about supplies and creature comforts, the über-rich are ready to hire small armies of guards to keep them safe from the plebeians. Rushkoff made the argument that the smartest investment was in long-term solidarity to tackle our collective challenges. They rolled their eyes. With their ideas of colonizing Mars and reversing the aging process, to the über-rich, the future of technology is about "escape from the rest of us."[74]

Mark Zuckerberg is building a $300 million compound in Hawaii that includes a five-thousand-square-foot "little shelter."[75] Ilya Sutskever, co-founder of OpenAI, told a group of researchers, "We're definitely going to build a bunker before we release AGI," so-called artificial general intelligence. "Of course," he added, "it's going to be optional whether you want to get into the bunker."[76]

In their mindset, winning involves doing whatever it takes to earn enough money to protect themselves from the damage that they are bringing about. But no man is an island; dreams of escaping death or apocalypse are nothing but dreams, and closer to nightmares at that. Those who wish to turn the tide of history through science and seek to control others don't realize how "suicidal this demand is," argued Popper.[77] Sooner or later, a boomerang effect ends up striking those who produced and profited from risk, warned Beck. Although it would be much less fulfilling for their egos, the ultrarich would be safer by doing a better job of building a more just world.

A metaphor from the comedian Jerry Seinfeld puts it well:

There are many things that you can point to as proof that the human is not smart. But my personal favorite would have to be that we needed to invent the helmet. . . . [Instead of avoiding danger,] we chose . . . to come up with some sort of device to help us continue enjoying our head-cracking lifestyles. . . . And even that didn't work because not enough people were wearing them so we had to come up with the helmet law. Which is even stupider, the idea behind the helmet law being to preserve a brain whose judgment is so poor, it does not even try to stop the cracking of the head it's in.[78]

When it comes to rule by prediction, we don't even have a helmet, much less a helmet law. And the head-cracking activities are mostly being orchestrated by superrich prophets who envision such a bleak future that they are building bunkers and hiring Navy SEALs for protection. We can do better. We deserve better.

Why would the new prophets engage in such destructive practices if they know they are bringing about havoc? Hannah Arendt argued that only destruction can satisfy the voracity of someone who cannot accumulate more power through wealth. The new prophets are already so rich that earning $10 billion more wouldn't feel like anything. And yet their lust for power is still there. So what do they do? They start destroying, because "the most radical and the only secure form of possession is destruction, for only what we have destroyed is safely and forever ours," writes Arendt.[79] We own what we destroy because never again can it belong to another. The logical result of endless power accumulation is self-destruction.[80]

"Power is not an instrument that its possessor can use with impunity," writes Robert Caro. "It is a drug that creates in the user a need for larger and larger dosages."[81] In the same vein, in her book *Autocracy, Inc.*, Anne Applebaum argues that authoritarian regimes in the twenty-first century are not bound by "ideology, but rather by a ruthless, single-minded determination to preserve their personal wealth and power."

THREADS GATHERED

The extensive use of machine learning is fueling the illusion of predictability. It's making society and the economy less volatile but more fragile. Big data can make it seem as if we were on the verge of attaining near omniscience with every extra data point we collect, but no amount of data will help us prepare for the unpredictable, for the times when the future does not resemble the past.

The extensive use of predictive AI is thus making society less resilient while seeming to make it stronger. By introducing more complexity into the system, and shrouding its fragility, we are making the world more unpredictable. By collecting more data than ever, we have achieved less predictability than ever.[82] The more we try to control the world, the more we create monstrous forms of uncontrollability—like nuclear and systemic risks. It's as if the very nature of existence rebels against being dominated by creating more extreme forms of uncontrollability.[83]

Unpredictability, arising from partial knowledge, social complexity and reflexivity, scientific and technological breakthroughs, flukes, and blindness, is never going away. Given this unsettling fact of life, how should we live?

PART THREE

RETHINKING PREDICTION

CHAPTER SEVEN

TRUTH, VIRTUE, AND BEAUTY

Why ideals outdo predictions

While the new prophets dominate the halls of power, effective altruists are whispering in the ears of the new prophets; they are the prophets of the prophets, the prophets of profit, the new generation of utilitarians. In Silicon Valley, effective altruism has become the "secular religion of the elites," writes Leif Wenar, professor of philosophy at Stanford University.[1] Although effective altruists are technically philosophers, in my book they land on the side of prophets, because one of their marks is that they are more interested in the future than the present.

When I was a PhD student in New York City, I was invited to an effective altruism party in Brooklyn, when the movement was in its very early days, around 2012. Effective altruism sprang out of the philosophy of Peter Singer, a Princeton utilitarian professor who famously argued that if it's in our power to prevent something bad from happening, and we wouldn't need to sacrifice too much to do it, then we should do it.

If you were walking in a park and saw a drowning child, you should (and would) save the child, even if that meant ruining your expensive shoes. And, if that's true, then you should be donating money to charity, because your money could save a child who is in dire straits halfway across the world. What does it matter that you live in different countries? If you can help, you should.

Effective altruists were a new young generation of philosophers who carried Singer's torch. Not only should you help, they argued, but you should do so effectively. Donating $100 to an ineffective charity will do much less good than donating the same amount to an effective one. And donating $100 to a charity in a part of the world where it can go a long way is much more effective than trying to save people yourself.

If you are feeling sympathetic toward those arguments, so was I, and so are most idealistic youngsters in their teens or twenties. Altruism done effectively—what's not to like?

I arrived at the Brooklyn brownstone by myself. I knew maybe two or three of the people who lived there, and they were encouraging me to open a chapter of effective altruism at my university. I was welcomed with a big smile by one of the hosts, who told me to hang my coat at the entrance, led me to the living room, and got me a drink. On a first glance the party seemed like any other party I'd ever been to; there were drinks and snacks, and people were talking and laughing. But something felt different.

Although there were a few people scattered around, the center of the party was a group of ten or fifteen guys, who were mostly deferring to one guy. A few other guys were talking in groups of two or three. And then it hit me. It's all guys. I looked for women around me, and spotted one after a while. I was used to being surrounded by men in class—that's what philosophy classes are often like, with numbers similar to those found in physics—but I'd never been to a party in which the gender ratio was so imbalanced. You could smell the testosterone.

I approached the main circle of conversation. The topic was earning to give. If you can do the most good by donating to charity, then you can do more good the more you donate. If that's true, then you should aim for the highest-paying career possible so you can donate the most possible over your lifetime. You can save more lives by getting a well-paid job in the city and donating a significant part of your income to charity than by being a doctor or a teacher in a deprived area. And that's how effective altruists came to encourage well-meaning undergraduates to become bankers.

In the conversation I was a witness to, guys were being encouraged to brag about how much they earned and how much they were donating to charity. I winced, but made an effort to remain open-minded. I knew the thinking behind this distasteful practice. The idea was that if you do good and stay quiet about it, you don't maximize the good, because if you talked about it, you could inspire other people to do the same. After all, human beings are social creatures. It still felt like one-upmanship, though.

One guy said that he wanted to call attention to the amazing donation another young man had made after going into finance. There was a wave of expressions of approval and congratulations, some of them visibly steeped in envy. The finance guy, tall, with dark hair and a boyish look, ducked his head slightly and waved a dismissive hand while a barely suppressed self-satisfied smile tugged at the corners of his mouth, his shoulders rising slightly as he basked in the praise. Keep an open mind, I repeated to myself, trying to suffocate the screaming Aristotelian in my mind who prefers to be in social circles that value virtues like modesty.

Even back then, effective altruism felt cultish, like a religious group with devotees discussing the meaning of passages in a sacred text. It wasn't only that their social experiment included living together in a house; it's how hierarchical it seemed, with a clear leader orchestrating the very philosophical conversation and establishing the doctrine.

The next topic was the risk of getting corrupted by earning huge amounts of money. Wouldn't it change you, becoming a banker? The consensus was that it wouldn't—not if you had gone into that career with the intention of donating money to charity; not if you stayed within a group that would steer you in the right direction. If you donated enough, you'd never get used to a luxurious lifestyle. And that's why people were encouraged to sign a pledge, to tie your fifty-year-old self to the commitments of your twenty-year-old self.

The excitement was palpable, the hubris unbridled. This group would go on to do great things for the world. "Could you imagine what a tragedy it would be for the world if something were to happen to this building tonight?" one of the boys said. I looked around me expecting laughter but found nods instead. That was my cue. I quietly grabbed my coat and made my way out, knowing that the group of self-appointed Messiahs were not who I would be spending my evenings with. That, however, would not be the last I would hear from them.

"Advising billionaires on how to give away their money and encourage them to give more is definitely not where I saw my life going," said William MacAskill, perhaps the best-known figure of effective altruism, and my colleague at Oxford until recently.[2]

Elon Musk has promoted the books of effective altruists on social media and has said that effective altruism aligns closely with his beliefs.[3] He has even spoken at effective altruism events. Musk has repeatedly justified some of his actions, like his feud with OpenAI, by referring to effective altruism arguments of wanting to save the world from existential risk. OpenAI and other tech companies have also become populated with effective altruists, and the ideology has been blamed for the temporary firing of CEO Sam Altman.[4]

It makes sense for the two groups of prophets to join forces. The first utilitarians (the predecessors of effective altruists) were the allies of the prophets of the nineteenth century—the social scientists and the bureaucrats. Effective altruism justifies the new prophets' concentration of wealth and power. It gives tech prophets the moral permission to do what they want and feel like superheroes at once, even when the rest of the world might look to their actions in horror. For effective altruists, that tech executives seem highly questionable to ordinary people is nothing to worry about, because they argue that people are too emotional and irrational, just simply wrong, and it's normal for ethical implications to be counterintuitive. Upon reflection, it's more of a put-down than an explanation—perhaps because put-downs are so much more effective than explanations.

Crucially, the success of both the new prophets and the effective altruists depends on our faith in prediction. But there are good reasons to limit the reach of prophecy, to favor right over might, truth over power, and ideals over predictions.

REASONS NOT TO BE AN EFFECTIVE ALTRUIST OR A UTILITARIAN

While effective altruists have been quite successful in exporting their views to the world outside academic philosophy, few have acted as a counterweight, exporting the criticisms that are well known within philosophical circles.[5] What follows is my attempt.

Effective altruists are the latest iteration of an older brand of ethics called utilitarianism. Even if you have never studied utilitarianism before, there's a good chance it comes naturally to you when thinking about moral issues, partly because utilitarianism is superficially intuitive (though it has counterintuitive implications), and partly because the predecessors of effective altruists were remarkably successful in converting policymakers to their views.

Consequentialism is an ethical theory that evaluates actions based on their consequences. Utilitarianism is a form of consequentialism that aims to maximize utility (usually defined as well-being or happiness). It judges the morality of actions based on their likely consequences for happiness and suffering. Doing the right thing amounts to maximizing well-being for the many.

According to some of the latest publications, effective altruism claims it's not the same thing as utilitarianism because it doesn't argue that one must sacrifice one's own interests if one can benefit others; it doesn't claim that the ends always justify the means; and it doesn't claim that the good equals the sum total of well-being. It seems that at least some of these tenets veer away from effective altruism's origins, and as we will see below, there might be reason to be somewhat skeptical of what they claim. In any case, many effective altruists are certainly also utilitarians, the movement was born out of utilitarianism, and the criticisms below apply to both.

Human Sacrifices

Staunch utilitarians or effective altruists might think they do a great good for the world, but I wouldn't recommend them as companions. They are not very good friends, neighbors, family members, or even doctors, because loyalty, justice, or rights aren't concepts that square easily with their worldview. That means that if you're in their vicinity, and it becomes convenient (for the greater good, naturally) to throw you under the bus or in front of a train, they will. They also think the work they do is so important that it trumps most other things.[6]

Derek Parfit was a philosophy professor at Oxford, and one of the grandfathers of the movement. Richard Jenkyns, a professor emeritus in classics who was Parfit's colleague at All Souls, said, "Would I have gone into the jungle with Derek? No, I would not. He would have found a morally and logically compelling reason why the jaguar should eat me rather than him. The loss to philosophy if he were the one consumed would be unthinkable."[7] I wouldn't want to find myself in a jungle with an effective altruist either.

According to a (modified) classic example meant to criticize utilitarianism, five patients lie sick in a hospital awaiting an organ transplant. A homeless person walks into the hospital with a broken finger; other than that, he is perfectly healthy. The homeless person doesn't have a family, and he has a reputation for being nasty. It occurs to the doctor that no one would miss him, and that she could take his organs and save the other five patients.[8]

If the doctor were a utilitarian, she'd kill the healthy patient to save the five. That's the effective thing to do, because five lives outweigh one. Some of the sacrifices that utilitarian types are willing to make seem to me as barbaric as the human sacrifices of ancient people like the Aztecs.

You might think that the doctor example doesn't even work under a utilitarian framework because it would create bad consequences in the long term. If people couldn't trust their doctor to treat them, they would end up not going to the doctor at all, and that would be bad for public health. But what if the doctor managed to keep it a secret?

Secrecy and Insincerity

Derek Parfit thought that the utilitarian Henry Sidgwick's *Methods of Ethics* was "the greatest work of moral philosophy ever written" and that it contained "the largest number of true and important claims" of any book in the history of ethics.[9]

Sometimes, argued Sidgwick, secrecy can "render an action right which would not otherwise be so." He goes on to suggest that it's reasonable for a utilitarian to desire that "the vulgar should keep aloof from his system."[10] What Sidgwick suggests here is that it can be alright to have an elite group of utilitarians making the important decisions on behalf of the rest of society. While everyone else should play by the book, those at the top are allowed to break the rules in secret for the greater good. If effective altruists subscribed to utilitarianism, as they seemed to do a few years back, they wouldn't necessarily confess to it, and they would feel perfectly morally justified in keeping that, and other kinds of rule breaking, secret—as long as it maximizes the good.

Not only will the utilitarian be willing to throw you at the jaguar if necessary, but he won't even be forthright about it. You can never trust a utilitarian because their goal is not to share the truth with you so you can make a decision for yourself. That would be a Kantian respecting your agency (or autonomy, as we say in philosophy). A utilitarian will tell you whatever he needs to tell you to maximize utility. They might be driving you to the slaughterhouse while telling you they're taking you someplace nice, because that'll save you the anguish. Or they might tell you they don't think that ends justify the means, because that way you'll be more likely to donate money to them so they can do with it what they think will do the most good. With a utilitarian, and with an effective altruist, you never know whether they're being sincere or whether they're just polishing their public relations campaign.

Disregard for Autonomy

Utilitarians and effective altruists are fond of playing gods. When *they* "save" a child's life because they donate money, they ascribe hero status to themselves without acknowledging the role of health-care workers or the parents of that child. It would be more accurate to say that donating money can help others save lives, but that sounds much less impressive.

Autonomy is both the ability and the right to govern your life; it's your agency. As an adult human being, you are capable of deciding what your values are, what is meaningful to you, and of acting in accordance with those values.[11] You make an autonomous decision when you fully own it, when it is a choice that you can endorse upon reflection. If you don't have autonomy, you don't have freedom, because your life is decided and controlled by others. Autonomy amounts to having power over your own life.

Respect for autonomy is at the foundation of medical ethics. It's why doctors have to ask for your consent before performing any invasive procedure on you. It's also the bedrock for human rights. But it doesn't play a role in utilitarianism above and beyond its possible contribution to utility. That is, having control over our lives tends to contribute to our well-being. But if a utilitarian thinks he knows best what to do about your life because he's come up with numbers and has a degree from Oxford, he'll be happy to decide for you and take that control away from you. To justify his knowing best, he has to make some doubtful assumptions.

Questionable Assumptions

The first questionable assumption is that we can clearly define and measure utility. Some people define it in terms of pleasure, and others think about it as happiness or well-being. While pleasure is undoubtedly an important component of the good life, it doesn't quite cover everything we value. We care about things like truth, love, and justice, even when they bring us no pleasure. Would you rather have a spouse who continually cheats on you but is good at hiding it (thereby maximizing their pleasure and yours), or would you rather know about it even if the truth will cause you pain? Many people would rather know.

Robert Nozick came up with a thought experiment called *the experience machine* to show that most people care about things other than pleasure. Imagine you had the option to be connected to a machine that could simulate reality in your brain. It could create a wonderful life for you, full of friends, and social recognition, and amazing experiences like winning an Olympic gold medal. It would all be fake, of course, but it would feel real to you. (You wouldn't remember being connected to the machine.)

Most people would prefer to have real friends than imagined ones, even if imagined friends might give you more pleasure and wouldn't do annoying things like challenge your views from time to time, or get an irritating partner, or forget your birthday.

Although I share the utilitarian desire to minimize pain, the longer I live, the more I think that pleasure and happiness are overrated, and that trying to maximize them is about the worst way to go about life. Happiness is so elusive that it tends to evaporate when we seek it. A utilitarian who wants to have a friend because he's read the literature on friendship and it seems to be good for happiness and health will miss out on reaping those benefits, because you only get them when you pursue friendship for its own sake—when you care about people because you care about *them,* not out of selfish motivations.

A further dubious assumption the utilitarian makes is that we have enough predictive abilities to see into the future. To be a utilitarian or an effective altruist, you need to have a good sense of where different courses of action lead to compare the consequences and choose the one that leads to the most good.

Effective altruists argue that we should follow expected utility theory, which states that rational decision-makers should choose actions that maximize expected value, calculated by multiplying each possible outcome's utility by its probability and summing these products. That means that doing something that has a 10 percent chance of saving a hundred lives is better than saving five lives for sure, because the expectation is that ten lives will be saved in the first case, and ten lives are twice as many as five lives. But that presupposes we can anticipate the possible consequences of our choices and estimate how likely each outcome is to occur.

Another question that arises with this kind of calculation is: When do we stop counting? Every course of action leads to a first wave of consequences, which in turn creates a second wave of consequences, and so on, until the end of life on Earth. Should we compare actions on the basis of their expected first-wave consequences? What if the second-wave consequences are weightier? You see the problem, not to mention the constant effect of the unforeseeable. While the predictable can result in positive expected value, the unforeseeable can blow up in our faces in the form of unexpected disvalue.

Unexpected Disvalue

In 2006, Holden Karnofsky and Elie Hassenfeld, hedge-fund analysts in their twenties who were inspired by effective altruism, quit their jobs and founded GiveWell. Neither of them had experience in aid, but they knew numbers. For altruists to give more effectively, they thought, they needed more data, and they could provide it. GiveWell would do research on aid and recommend the best charities.

Their research suggested that one of the most effective ways to save a life was to donate to charities focused on preventing malaria. They calculated that you could save a life by donating $820 to a charity that gave out bed nets, protecting people from the mosquitoes that carry the disease.

The world is always richer than we suspect. It's rebellious too. It refuses to bow to our calculations, our simplifications, our predictions. There is more than one way to use an insecticide-treated net. Had our effective altruists lived in Madagascar, it might have occurred to them that they are also great for fishing. In 2015, *The New York Times* reported that malaria nets were contributing to overfishing, which in turn was endangering fragile food supplies across Africa.[12] These harms were conveniently omitted from the calculations of effective altruists.

One of the top charities that GiveWell recommends contributes to having children vaccinated by sending money to dangerous regions in northern Nigeria. Armed men have attacked locations where this money is kept, in one case killing two people and kidnapping two children while searching for the money. These deaths and harms don't appear in the calculations either, and we're not sure how often these kinds of incidents happen.

As Leif Wenar points out, aid can have negative effects. And these effects are somewhat unpredictable. When charities hire health workers who previously worked in public institutions, infant mortality rises. One hypothesis to explain this unexpected negative effect is that health workers save more lives in public institutions than through charities because what is most needed are interventions in cases of medical emergencies, and when people have a medical emergency, they go to public institutions, not charities. Giving deworming medicine to children at school without the consent or knowledge of parents can make people feel distrustful of authorities and of aid. Even worse, after years of effective altruists recommending deworming as one of the best ways to do the most good, evidence now suggests that there are no benefits from mass deworming. GiveWell no longer recommends it.

GiveWell's website claims that the $2 billion donations in aid they have secured since 2009 will have saved more than 200,000 lives, but it doesn't account for the deaths. Tackling extreme poverty, argues

Wenar, is partly about shifting power to the worst off. And some of the "help" of effective altruists has done nothing of the sort.

Effective Altruism Meets Tech Money

At first, effective altruists were focused on alleviating global poverty. But most millionaires and billionaires are not interested in alleviating poverty; if they were, they probably wouldn't be as wealthy. The rich also suffer from the halo effect of thinking they became rich for reasons having nothing to do with luck. William MacAskill said he tried to talk with Elon Musk about poverty and "got little interest."[13] Lucky for the effective altruists, they stumbled upon a topic that they would turn into a gold mine: AI safety.

Nick Bostrom, the former director of the now-defunct Future of Humanity Institute at Oxford, published *Superintelligence* in 2014.[14] In it, he warned of the dangers of building too powerful an AI. The Uehiro Centre for Practical Ethics, where I worked then, shared a building with the institute, so I heard and saw things here and there. "Do you know what made the book so successful?" a staff member of the institute at the time asked me. I shook my head. "One tweet from Elon Musk promoting the book; that's what turned it into a bestseller." Musk went on to donate £1 million to the institute.

It was around this time that rumors started circulating about polyamory in effective altruism. It was a way to maximize utility, a more efficient and less impartial allocation of emotional resources, a rational examination of social norms. To each their own. But what was concerning was that the women tended to be younger, that the personal relationships were mixed with professional ones, and that the men held power over professional jobs and opportunities. Women were being used as just another source of expected value, yet another human sacrifice to the god of pleasure.

It was shocking to witness in real time how a boys' club gets formed and institutionalized. When I was in my twenties, I thought that my generation was going to overcome most forms of sexism. But as far as I could tell, the only difference between this boys' club and previous ones was that this one had found a purported ethical front to ratio-nalize their actions. They had better public relations and they were better at disguising the sexism, but they were still placing each other's male buddies in positions of power, justifying it because those bud-dies would be more altruistic than alternatives, and then using their power to create small harems.

In the summer of 2015, one of the effective altruism organizations got accepted into Y Combinator, the prestigious start-up incuba-tor, which further opened doors to the pockets and halls of the tech executives.

Soon, many more millions started pouring into effective altruism institutions, and they weren't spent on alleviating poverty. About $23 million went to buy Wytham Abbey, a twenty-seven-bedroom historic mansion near Oxford.[15] Another $3.5 million went to buy Chateau Hostačov, a Renaissance castle in the Czech Republic. That money came from the FTX Foundation, the FTX crypto exchange's philanthropic arm; we'll get to that.[16]

For his latest book, William MacAskill allegedly paid $12,000 per month to one of the PR firms he hired and mentioned a maximum budget of $10 million for PR. The funds came from Dustin Mos-kovitz, a co-founder of Facebook.[17] You don't usually see that kind of money in the publishing world; the budget was so high that it prompted someone in the PR firm to blow the whistle. The release of the book was celebrated at a luxury restaurant where the tasting menu cost $438 per person.[18] The money and high-powered connec-tions worked; interviews with Trevor Noah, Lex Fridman, and Ezra Klein and a profile in *The New Yorker* were among the PR successes. Not coincidentally, the title of the *New Yorker* piece on its website refers to MacAskill as a prophet.

How could effective altruists justify such spectacular expenditures? In not too different a way from how project managers in chapter two multiplied the rate of seagulls' consumption of grasshoppers by the value of grain eaten by each grasshopper to justify building a reser-voir. You can always make the numbers work for you if you're clever enough, but good metrics don't equal good morals. Numbers cannot tell you what the right thing to do is.

Infinity

The clever effective altruists learned a nifty mathematical trick: using infinity. If you're trying to prevent something infinitely bad from happening (like human extinction), or advancing something infinitely good, then *anything* can be justified.

That's how longtermism came into the picture. The main idea is that, just as distance across space isn't morally relevant (a child on the other side of the world is just as morally valuable as the child next door), distance from the present in time isn't relevant either, and someone living in the future has as much of a moral claim to any resources as you do. Like their arguments about poverty, this idea sounds deceptively straightforward and benevolent on a first read. But its implications are grim.

Let's say that there will live eighty trillion people in the future of humanity; it's a number that effective altruists have floated around, but like many of their numbers it's more than a little questionable, especially in light of just how inaccurate past population forecasts have turned out to be. Some estimates include such dubious assumptions as interstellar travel and the uploading of human minds to computers; I'm not kidding. Anyway, let's just roll with eighty trillion. The world population today is around eight billion people. Just as the lives of five patients trump the life of one, the lives of eighty *trillion* people undoubtedly trump the lives of eight billion.

If you spend all your time and money reducing existential risk by even one billionth of a billionth of a percentage point, that is still billions of times more expected value than saving the lives of billions of people now. In other words, it's much better to work to reduce the hypothetical risks of hypothetical future people than to help living, breathing, non-hypothetical people.[19]

Roger Crisp, professor of moral philosophy, welcomes me in his office at St. Anne's College in Oxford. Roger was one of my PhD advisers, and years ago we spent many hours in this office discussing my thesis. "It looks the same," I say, sitting down on the long couch. "A few months ago, the ceiling collapsed. If you had been sitting there at the time, it would've killed you," Crisp says. I look up at the recently painted ceiling and resist the urge to move. When I was a student, part of the magnificent wooden ceiling in the Great Hall at Christ Church collapsed in the middle of the night; it could've killed someone too.

"Once your scope is infinite, you can do anything, because nothing makes any difference," Crisp says. "There is an infinite amount of good stuff, and an infinite amount of bad stuff, and if you do something bad, you're not going to create more bad stuff, because there is already an infinite amount of it." In the vastness of infinity, there is no making the world a better place; there is no making a difference one way or another. Every act, however large its impact, gets dwarfed by infinity. You could murder someone in cold blood, but what is that suffering in the grand scheme of the infinite? Any amount of harm caused is just a blip in an ocean of forever.

Charles Dickens, a contemporary of John Stuart Mill's, knew utilitarians well. Seemingly concerned with the welfare of humanity and with "facts, facts, facts" ("data" is the contemporary term), the "eminently practical" utilitarians end up losing their humanity, straying from what matters most, much like Thomas Gradgrind in the novel *Hard Times*.

Even Peter Singer, one of the grandfathers of effective altruism and the most famous living utilitarian, has expressed his concern for the ideology of longtermism and the focus on existential risk. Having too broad a perspective in time shrinks our current problems and fellow human beings to specks of dust, and can provide a rationale for doing anything to maximize utility. In this sense, effective altruism shares important traits with Communism. It was the justification of advancing toward the (supposedly inevitable) long-term goal of human history that justified atrocities in the Soviet states. Similarly, Nazis used the goal of a "Thousand-Year Reich" to enslave and exterminate those who didn't fit in the plan.[20]

The risk of something infinitely bad also justifies outrageous recommendations like having a police state to prevent people from using technology for evil.[21] The kinds of policies advanced by some philosophers are analogous to living in a bunker because you're afraid of pandemics and war; you can still break your neck slipping in your home bathroom, though, not to mention the quality-of-life cost of living in a bunker. Police states with the kind of surveillance architecture we have built are much more dangerous than any AI.

Infinite Human Sacrifices

Utilitarianism has a weakness for infinite cases or numbers large enough to warp our intuitions. A famous one by Derek Parfit is the *repugnant conclusion*. Compare two possible worlds. In one, there are eight billion people, all with very high quality of life; there's no poverty, racism, sexism, wars, or chronic diseases. In another world, there are eighty trillion people whose lives are barely worth living.

The repugnant conclusion is that the latter is a better world because overall there is more utility in it: The sum total of well-being is higher because of the very high population number. Parfit was disturbed by this conclusion, and yet willing to "bite the bullet," as philosophers put it. To me, that's like driving off a cliff because Google Maps says your destination lies that way. (These days, you don't have to physically throw prophets off cliffs; just give them a maps app and they might do it themselves.)

In a classic example, there is a concert in progress that is set to last for an hour, and millions of fans are watching it live, each one relishing every minute of bliss-filled music. But one of the staff has suffered an accident with the electrical gear, and we cannot rescue them without turning off the equipment for fifteen minutes. They are receiving extremely painful electrical shocks. As long as fans are enjoying the concert enough, however, utilitarianism would recommend that the show go on. The utilitarian argument is that the aggregate of well-being is more important than the well-being of any individual. Human sacrifices all around.

Here's another: the *utility monster*, created by the philosopher Robert Nozick to criticize utilitarianism. Imagine that there is a person who draws inordinate amounts of pleasure from things; if chocolate gives one unit of pleasure to you, it gives a thousand units of pleasure to them. Utilitarianism would dictate that we give the chocolate to them instead of you. That kind of math gets you to ugly places. I enjoy math; had I not studied philosophy, I might've studied mathematics, but there is a limit to how much numbers can help in ethics.

Another kind of utility monster is the one who *creates* much more utility than others. Effective altruists have argued that saving a life in a rich country is more important than saving a life in a poor country because richer countries have more innovation and are more economically productive.[22]

Yet another argument in the tool kit of effective altruism is the idea of *replaceability*. If you don't take the job as a banker, someone else will, and they won't do as much good. The most troublesome side of the replaceability argument is that it has no moral limit to what it can justify: from working as a drug dealer to dropping a nuclear bomb. As long as someone equally qualified as you is willing to do the job, and you think of yourself as someone altruistically minded who will do less harm (or more good) than someone else, anything goes—no job is off limits.[23] Worse, you should probably take the most harmful job you can think of, because otherwise your replacement is almost guaranteed to do more harm than you. The argument also doesn't consider the possibility that someone else might *not* take the job. It doesn't give humanity the benefit of the doubt.

Elon Musk has used a similar argument to build his own AI. If he doesn't do it, others will, and he claims he can do it better and save the world.

Given utilitarianism's level of comfort with human sacrifices, it wasn't surprising when reports of sexual harassment and misconduct in the circles of effective altruism came to light. One of the victims commented that she was strongly discouraged from going to the police to report the incident; "the accused man's important career was a focus of the conversations," she said.[24]

It makes me sad to think about young people being persuaded to sacrifice their one life to become bankers or tech employees for the money instead of encouraging them to follow their dreams as doctors, artists, or whatever else suited them. I remember one instance of a young guy asking an effective altruist, "But what about someone like Gandhi?" The effective altruist responded, "Yeah, but what are the chances that you'll become someone like Gandhi? You have much better chances as a banker."

Occasionally an effective altruist will express doubts about their stance. "What if we're just wrong? What if AI is just a distraction? . . . It's very, very easy to be totally mistaken."[25] But is this seeming ray of humility just another PR stunt? They act with the certainty of prophets, and through the tech prophets they are commanding billions of dollars, banking on questionable theories applied to even more questionable predictions, treating people like pawns.

Even though effective altruism brands itself as an open-minded movement, sensitive to empirical data, hyperrational, and welcoming of criticism, the reality seems different from the image portrayed. When criticism ensued from within, members were told that it was dangerous to go against the main tenets and figures of the movement; it could hurt donations that were saving lives. For a utilitarian, failing to save a life is just as bad as killing someone. No sane person wants to be responsible for a death, so everyone fell into line and kept quiet.

But perhaps the most exemplary story to showcase the problems with these ideologies is that of Sam Bankman-Fried.

More Unexpected Disvalue

In 2022, effective altruists justified their lavish expenses, like buying two castles, by arguing that they were an investment to recruit the "next Sam Bankman-Fried," then a wunderkind of crypto who became effective altruism's biggest donor. Around three months later, he became their biggest PR mistake.

Ten years earlier, when Sam Bankman-Fried was a student at MIT, he attended a talk by William MacAskill on "earning to give" and went to lunch with him. Bankman-Fried was sold on both effective altruism and the idea of earning to give. After he graduated with a degree in physics two years later, he worked first for a trading firm, then briefly as director of development at the Centre for Effective Altruism, and then went on to found his own trading firm, Alameda Research. In 2019, he also founded FTX, a cryptocurrency derivatives exchange. In 2021, he was recognized by *Forbes* as the richest person under thirty in the world. He was twenty-nine years old and worth at least $22.5 billion.[26]

One year later, the house of cards came crashing down. FTX collapsed amid revelations of massive fraud and misuse of customer funds. Bankman-Fried was arrested, tried, and convicted on seven criminal counts including fraud and money laundering. He had been secretly taking FTX customer deposits and transferring them to his trading firm Alameda Research, where the money was used for trading, personal investments, loans to executives, political donations, and luxury real estate.

The wunderkind thought he would be able to get away with it all because he was making bets that had a positive expected value. Had bets turned out consistently well, customers wouldn't have wanted their money back and no one would've discovered the missing money. But *expected* value is not value; it's a gamble, and sometimes bets don't land your way.

Bankman-Fried was so committed to making decisions with a positive expected value that he claimed he was willing to flip a coin that, if it landed tails, would destroy the universe, but, if it landed heads, would make the universe slightly more than twice as good as it is now (2.00000001x). He claimed he would be willing to flip that coin *repeatedly*. In a way, he did flip a coin similar to that one, with his own business and the money of his customers. The wunderkind turned into a convicted fraudster sentenced to twenty-five years in prison.

Michael Lewis called his book about the topic *Going Infinite*, referring to Bankman-Fried's use for and desire to earn infinity dollars; the book starts with a quotation from the mathematician David Hilbert about how infinity is nowhere to be found in reality. If we make our moral decisions on the basis of the infinite or near infinite, or on predictions far away from the present, we stray from truth and what matters.

In the wake of Bankman-Fried's spectacular fall, effective altruists tried to distance themselves from their poster child. They condemned his actions and said that if someone had engaged in deception and fraud, then they had abandoned the principles of effective altruism. But I'm not sure any of it is credible. As long as Bankman-Fried was donating enough money to the causes with the most expected value, then he was doing the most good. His main mistake, from the perspective of both effective altruism and utilitarianism, was getting caught.

For a non-utilitarian, Bankman-Fried lacked nuance in his understanding of morality. He stopped reading books when he was around eighteen years old, replacing them with utilitarian message boards online. "I would never read a book," he said. "I think, if you wrote a book, you fucked up, and it should have been a six-paragraph blog post."[27]

Maybe Bankman-Fried would've ended in a similar situation if he had never come across effective altruism. But maybe he would have never attempted to become as rich as possible had he not come across the grand idea of earning to give. In Michael Lewis's assessment, the "big changes" in Sam's life "tended to come after some other person had made an argument to him that he couldn't refute." Perhaps if Bankman-Fried had had a spectacular teacher in high school who had taught him Dickens and a love for good books, his life would've gone better.

Overlooking Tragedy

One of the most powerful criticisms leveled against utilitarianism was articulated by the philosopher Bernard Williams, who argued that utilitarianism is a moral theory that doesn't only allow monstrous acts to be committed; it *requires* them. And that can't be right. Utilitarianism pushed Bankman-Fried to commit fraud.

Suppose you find yourself in a hostage situation in which the criminal in charge has lined up twenty people against a wall and is ready to kill them. The murderer makes you an offer: If you're willing to shoot one of them yourself, he'll let everyone else go.

Utilitarianism would require you to kill a human being. Not only that, but it makes it seem as if it were the obvious choice, as if it were a no-brainer. For utilitarianism, you're just doing the right thing. But a situation in which you kill a human being is not doing what's right; at best, it's doing what's least wrong. Utilitarianism cannot reflect that nuance.

Utilitarianism's great allure is that it can reduce any gut-wrenching situation to a calculation. If you can compare numbers, you can be ethical. Conflict in utilitarianism is impossible, tragedy is always averted.[28] At least on paper. But that is not an accurate reflection of the moral reality of tragedy on the ground.

Doing accounting with people's lives is sometimes making things too easy, as in the case of the murderer, and sometimes making things too hard. As Williams famously put it, someone who tries to justify the impulse to save the life of a spouse over that of a stranger has had "one thought too many." It's your life partner. You should save them without giving it a second thought, and if you do give it a second thought to do moral accounting, well . . . you've had one thought too many.

One way to present the criticism is that utilitarianism engages in a kind of category mistake: It treats moral matters inappropriately in the form of accounting or gambling, and in doing so distorts the moral landscape in ways that veil tragedy, obliterate justice, and turn flesh-and-blood human beings into utility units.

But for the purposes of this book, the most important objection to effective altruism and utilitarianism is that they both heavily rely on our predictive abilities to make decisions. If you temper your views about how far into the future anyone is able to peer, effective altruism loses much of its luster.

Seeing into the Future

Malaria nets weren't as straightforward as they seemed, deworming turned out to be as ineffective as it gets, and Sam Bankman-Fried has become a cautionary tale. The predictions of effective altruists have been spectacularly wrong even in a time span of a decade or so.

When Nate Silver asked William MacAskill whether the latter should have seen the collapse of FTX coming, MacAskill said it was unpredictable and cited a forecast from a prediction market site called Metaculus that estimated the chance of FTX defaulting on any customer deposit in 2022 to be about 1.3 percent.[29]

That's a perfect example of the inadequacy of too subjective a view on probability. Once you know about how Bankman-Fried had taken FTX customer deposits and was using them to make bets through

Alameda Research, you realize that the actual chance of FTX collapsing was much higher. Effective altruists' will to power led them to be conveniently naive, at best: They wanted the world to be a certain way; they wanted the future to bow to their beliefs; they believed an unrealistic forecast from a prediction market because it was in their interest.

These days effective altruists support causes based on predictions of what the world will look like, not in three months, not in a decade, but in hundreds and *thousands* of years. If predictions are never facts, the further away in the future they prophesy, the more they veer away from the world of facts. To sacrifice living breathing people for hypothetical future people thousands of years away is not so different from ancient tribes sacrificing human beings and offering their hearts to the gods. Neither the ancient gods nor the future that effective altruists envision a thousand years from now exists.

"If nerdy young adults see that they can get money or status or sex by intoning on Very Important Things while pulling numbers out of their pants, at least some will start pulling," writes Leif Wenar.[30] Young people with a taste for mathematics express being attracted toward the "rigor" of effective altruism and utilitarianism, but playing with numbers doesn't make a moral theory any more rigorous.

Utilitarianism in the Age of AI

Utilitarianism has always been a terrifying theory. It's never been a good idea to put yourself in the hands of someone who is color-blind to justice. But in the age of AI, it's all the more terrifying. Most human beings have psychological barriers that can moderate unreasonable views. A utilitarian might cave to spending an afternoon consoling a grieving friend instead of making money; they might consider it a moral failing, while the rest of us can feel relief to catch a glimpse of their humanity.

Peter Singer is a utilitarian, but he shows partiality to his family and friends; he has funded his children's education instead of donating that money to the worst off, for instance. He also has a reputation within the profession of being honest and generous with his time. "Do you genuinely think that your partiality towards your loved ones is a moral failing?" I ask him. He pauses to think. "There is a sense in which I see it as a moral failing, yes. If I were a perfect moral being, a perfectly rational being, then I wouldn't be partial in that way, but then I would be a very different kind of being, less human. Human beings have evolved to be partial to those around us."

"Artificial intelligence hasn't evolved that way," I retort, "so should we turn it into a foolproof utilitarian?" "Yes, I think we should," Singer says, with one caveat: We shouldn't allow it to make Bankman-Fried-style bets with humanity.

AI doesn't have to look into the eyes of fellow human beings. It doesn't feel pain at the sight of others' pain. It can be programmed to be a perfect utilitarian—a moral monster. And it can scale to cover the span of every sentient being on the planet. For that reason alone, it has never been more important to reflect deeply about ethics.

In my experience, utilitarianism tends to be used as a justification by the elite and the powerful. In the battle between might and right, it tends to side with might. A small group of self-appointed enlightened ones make decisions for the rest of us in the name of the most good for the most people, caring little or nothing about what those people believe or how they want to live. It's no wonder that Bernard Williams suggested Sidgwick's nineteenth-century utilitarian views comported well with colonialism.[31]

VIRTUE, RIGHTS, AND JUSTICE: LIMITING PREDICTION IN ETHICS

In the face of ideologies that value hypothetical future risks to hypothetical future people as more important than the person suffering injustice right across the room, more embodied and situated approaches to morality value facts over predictions, and rights over calculations.

The term "deontology" comes from the Greek *deon,* or "that which is binding." When we talk about rights or principles, we are talking in deontological terms. Whereas utilitarianism focuses on consequences, deontology focuses on acting in accordance with duties, like respecting rights. Immanuel Kant is thought to be the figurehead for deontology. One of the pillars of his philosophy is respect for rational beings as ends in themselves, and never as a means to an end. People are not pawns, and we shouldn't be treated as objects. We have our own values, and we should be respected enough to be allowed to make decisions for ourselves. When you lie to someone, you are not giving them the respect they deserve, because you're not allowing them to make an informed decision.

Kant has often been ridiculed for the following example. Imagine you are with your friend in a cabin. A murderer knocks at the door. You open the door. The murderer asks where your friend is. Should you lie? No, argues Kant. Imagine you lie to the murderer: You say that your friend left through the back door. Unbeknownst to you, your friend had overheard the murderer asking for him and had in fact fled through the back door. The murderer then finds your friend and kills him. Kant argues that if you do something generally wrong, like lying, then you are liable for whatever unforeseen consequences might ensue. That is, we might be tempted to make exceptions to the rule against lying because we think the consequences of telling the truth will be bad, but we can never be certain of what the consequences will be. If you do your duty, then whatever might go wrong is not on you.

I think you should lie to the murderer, because the consequences are proximate enough that you stand a decent chance of predicting them accurately. Kant's argument has been misinterpreted to mean that you shouldn't lie because you'd wrong the murderer, but that's not his point at all. For decisions in which the consequence isn't immediate and where many interacting factors might come into play, we shouldn't be so confident in our ability to predict the future that we'd be willing to do something wrong for the sake of an expected outcome that might never come about. Kant's criticism of utilitarianism can be understood as a criticism against relying on prediction in ethics.

One of the drawbacks of utilitarianism is that it makes us vulnerable to moral luck. For a utilitarian, if you make a decision that ends up having bad consequences due to unforeseeable factors, you did something morally incorrect, irrespective of your intentions. Conversely, if you try to kill someone but through a lucky set of happenings end up saving their life, you're a moral hero, regardless of your intentions. But we tend to think about the realm of morality as one that is subject to responsibility and autonomy, not luck. We tend to think that we shouldn't be morally judged for what we can't control and we're not at fault for.

For utilitarianism, the building blocks of morality are consequences. For deontology, it's duties. For virtue ethics, it's about having a good character. Under deontology and virtue ethics, the most common ethical competitors to utilitarianism, moral agents are better shielded from moral luck. You do what's right. If other people don't, that's on them. If a fluke or some unforeseen domino effect ends up creating a bad state of affairs, as long as you did your duty, as long as you were virtuous and not reckless, that's not on you.

One way to interpret deontology is through the lens of heuristics. Heuristics, remember, are condensed knowledge. Respecting rights and cultivating virtues like honesty, bravery, and generosity tend to work better in the long run than maximizing consequences. We know this through millennia of experiences woven into our culture; that's why your parents taught you not to lie. Heuristics can beat prediction not only in science and economics but also in ethics.

The ultimate irony is that even if we agree with the utilitarian aspiration of achieving the most good for the most people, we're more likely to get there in a world in which people strive for virtue, the respect of rights, and justice. As Bernard Williams puts it, even if utilitarianism were true, the world would be better if people didn't believe in it. And that's even more so if utilitarianism is false.

Utilitarianism is especially bad at securing justice. And justice is one of the cornerstones of any liberal democracy. In another classic example, a judge must decide whether to condemn an innocent person she knows to be innocent to prevent riots that are bound to cause many deaths. Utilitarianism would require that she punish the innocent person to prevent unnecessary deaths. That she is innocent is irrelevant to the utilitarian.

A disregard for justice is also the result of aggregating pleasure or happiness. The philosopher John Rawls famously argued that such an approach treats human beings as if they belonged to one body. If you had to consent to a doctor cutting your gangrened toe to save your life, that's one thing. But sacrificing one member of a group for the benefit of the group is not analogous. Utilitarianism doesn't acknowledge, much less respect, the separateness of persons. A person is not a toe. A person has their own values, life plans, and objectives. That is precisely the function of rights: to protect individuals against anyone willing to sacrifice them for their own agenda.

A predictive approach to ethics is likewise inadequate for matters of justice, inside and outside the courtroom. In criminal contexts, merely statistical evidence isn't enough. Imagine a prison yard in which there are a hundred prisoners; ninety-nine of them form a riot and collectively kill a guard. Only one person refrains from the action, standing alone in a corner. We have video footage that tells us this much but doesn't have enough resolution to make out identifying signs.

One option is to pick one of the prisoners at random and let him take the blame for the murder; after all, the probability of his guilt is 99 percent. We could also convict all of them, since the probability of their guilt is likewise 99 percent. But, as philosophers have argued, that's not the right kind of evidence for a criminal conviction; we need *causal* evidence linking one particular person to a particular crime.[32]

A causal approach to human relationships also applies in matters of fairness outside the courtroom. A professor needs to read a student's term paper before assigning a grade, even if she has a high confidence of who will get an A-plus and who is going to fail. You might be sure that your colleague will leave his dirty coffee cup behind, but it would be inappropriate to send him an angry email about it until he does.

So why are we denying people loans, jobs, apartments, bail, and other opportunities on the basis of merely predictive evidence? Because it's profitable and efficient—nothing to do with justice. These practices are not data-driven; they are profit-driven. We don't even check to make sure that the elements tracked by machine learning algorithms are causally related to the opportunities we are denying, and not just spurious correlations. We are straying away from truth as a value worth striving for. But truth is a fundamental pillar of democracy, and democracy is good for personal freedom, innovation, and business.

TRUTH

We're running low on people like Socrates these days. Or maybe they don't attract enough attention in a time in which strongmen seem to be held as heroes. If there is a Socrates out there, or a Simone de Beauvoir, they are probably too busy writing emails or having very important meetings with administrators, jumping bureaucratic hoops.

After the fall of Bankman-Fried, effective altruists went a bit quieter. But they are not going away. Utilitarians have been around for three centuries, always at the side of policymakers, rulers, and prophets. That's why it matters that we criticize them, that we don't ignore the impact they've had in justifying the agenda of the most powerful prophets of our times.

"Does utilitarianism care about truth, above and beyond investigating what increases utility?" I ask Roger Crisp, my old professor at Oxford. "No," he says. "That bothers me," I say. He shrugs with a small chuckle, palms rotating upward, as if saying, Well, it is what it is. We break out in nervous laughter after a long silence.

If tenured academics don't use our freedom of speech to speak truth to power, we are not worth our salt, and we will end up losing that freedom, which is so hard to win, so easy to lose, and so important for society. I got an education thanks to public money from Spain, the United States, and the U.K.; my loyalty lies with the public. The job of academics is more important than ever in a world sliding steeply into bullshit. I'm glad I rejected the job I was offered to rubber-stamp a corporation's algorithmic projects. If I told you the salary I passed up, you might think me a sucker, but if I had taken it, I wouldn't have had the freedom of speech to write this book, and you wouldn't be reading it.

The Nobel Peace Prize winner Maria Ressa warns that a world devoid of a commitment to truth is "a world that's right for a dictator." She knows what she's talking about. As an investigative journalist working in the Philippines at the time, she witnessed and called out the lies of the dictator Rodrigo Duterte. She was arrested for it.[33] She is one of the bravest people I've ever had the honor to cross paths with.

Bullshitting the public is an essential part of the autocrat's playbook. When tyrants distort the truth, their main objective is not to deceive but to intimidate. Autocrats often tell lies that are purposefully shameless and outlandish. When no one calls them out, when people nod along, it shows just how powerful they are. The more unbelievable the falsehood, the more they have proven their power. Remember George Orwell's *1984*. If the autocrat can make 2 + 2 = 5 "true," then he can make anything true. George Orwell didn't make up these tactics; he lived them.

Gaslighting is a weapon of authoritarianism. In a reeducation camp in China, a teacher stands in front of her students. "It's time for an exercise," she says. She places two water bottles on the desk; one is empty and the other is full. "I say the full water bottle is full of water. I also think the empty bottle is full of water. What do you think?" One student answers, "Both bottles are full!" "Good!" answers the teacher.[34]

The autocrat doesn't care about whether people believe him; he cares about whether people obey him.[35] And he cares about creating complicity around him. If you see the truth and call him out for his bullshit, he'll crush you. If you stay quiet, you become his accomplice. Liars and con men can only thrive in a world in which everyone else is a liar and a fraud too, so they build the world around them in their own image.

If you stay quiet long enough, if you're afraid enough to censor your dangerous thoughts about what's true, then you start losing your grip on reality. The shameless tyrant sows shame in the person he bullshits.[36] You start depending on the tyrant to interpret the world, because what your own eyes and rationality tell you is being denied by an authority who owns the thread on which your life hangs.

Doublespeak is one of the most powerful weapons of the autocrat. China's constitution guarantees citizens "freedom of speech, of the press . . . and of demonstration." There is a "parliament," there are "elections," people talk about the "rule of law," and the country calls itself a "republic." Those words do not refer to what they refer to in democratic countries. But Chinese citizens grow up thinking they do.[37]

In *Politics,* Aristotle argued that to remain in power, the tyrant must humiliate his subjects and sow distrust among them. Trust and solidarity between citizens are the enemy of the autocrat.[38] And truth is necessary for both trust and solidarity. If the powerful who happen to be in charge for now are bullshitters or fortune tellers, it's all the more important that we be truth tellers.

Academia

"I would rather discover one true cause than gain the kingdom of Persia," said the Greek thinker Democritus. For centuries, universities have been one of the most important if not the most important guardians of truth in our societies, along with investigative journalism and the judiciary. At best, academia is a refuge for those who are more interested in truth than in power. While I've met the most despicable people I know in academia, I'd be giving you a false impression if I didn't clarify that I've also met some of the most admirable human beings there, and the good ones outnumber the bad. If we lose academia in spirit, we will have lost a crucial source of truth telling. It is not a coincidence that this book was written by a tenured professor in a university like Oxford, but there are fewer and fewer of us, and it's getting harder to do our jobs.

I have the privilege, and the duty, to spend a good proportion of my time thinking about the issues that matter most, like what is a good life and how do we get there, what are the pillars of democracy, how do we preserve freedom, what can we learn from others, and what crucial signals might we be missing among all the noise.

In China, the party buys the loyalty of university professors by paying them for not attracting negative attention, for not making critical comments.[39] In the United States and the U.K., it's big tech who is buying professors.

Most of my colleagues working in computer science and AI ethics on both sides of the Atlantic have received money to fund their research from big tech or work part time with big tech. The worst part is that we are cheap to buy. Our salaries are so low (especially in the U.K. and Europe) in comparison to the seven-figure salaries that are found in tech companies that silencing us through grants is a minor investment for big tech, with big payoffs.

Before you judge academics too harshly, let me give you a sense of what the incentives are like. In the 1970s or 1980s, it was perfectly possible to buy a house in a central neighborhood in Oxford with the salary of one academic. Today, *two* academics with tenured positions would struggle to buy the same house. I am told that the situation is similar in California, Boston, and New York City, at a minimum. As an academic, you may hold a prestigious job, but if you can't afford to live in the city your university is in, it starts feeling much less grand quite quickly.

There are some people who have family or personal needs that mean they cannot afford the very low salary academics earn before securing a permanent position. You have to be able to afford working as a junior academic. Given the inequalities that the world faces, you might not be willing to feel bad for academics. Fair. However, academics are a crucial counterweight to prophets and prophecies, so even if you have no sympathy for them, society has an interest in allowing them to do their jobs.

The teaching and administrative load can be so crushing that if you don't find a grant to relieve it, you might never get any research done. (And if you're too successful at getting grants, your university might ban you from applying for another one to force you to do more admin and teaching.) If you work on AI or related fields, big tech is your best bet. But Google is not going to give you a grant if you've written about how the business model based on personalized ads is toxic for democracy. If you're a computer scientist, you might need a relationship with a big tech company to get access to the kind of compute that you need for your work, and that means that if you criticize them, you jeopardize both your career inside academia and the alternatives.

In the U.K., the infamous Research Excellence Framework evaluates the impact of academic research. If you get a high assessment, your department gets more funding. One of the ways to achieve a high-impact grade is if a company cites your work. It doesn't count as impact if an academic cites your work, because an academic citation might not result in practical implementations. The incentive for someone like me is to side with companies.

If I were to write about how wonderful the theft of personal data is, how innovative the exploitation of gig workers, and how safe and reliable generative AI is, I'd stand an excellent chance of being cited by Google, Microsoft, OpenAI, and the like. My department would get more funding. I'd also be invited to give talks to the big tech companies, which would further aid my career.

Instead, I sometimes get invited to advise governments or the European Commission, which is an honor and a citizen duty I take very seriously, but also a challenge. I have to take a day off work to do so, and the work doesn't go away; it just gets pushed to the weekend. I sit on a panel in Brussels. Next to me, a representative from big tech is being paid a fortune to be there.

There are many ways to silence academics. You can threaten them. You can buy them off. Or you can bury them in administrative work and teaching. If you agree that a commitment to truth is crucial to democracy and justice, and if you can provide a counterweight to the influence of big tech, fund academics. Don't fund them centrally by donating to a university or a department; it'll get lost in the system. Buy the freedom of an academic whose work you think can change the world.

Diogenes of Sinope was a famous Greek philosopher in the fourth century BCE. One of the curiosities that made him renowned is that he lived in a barrel. He had taken a vow of poverty and, much like Socrates, spent his time asking uncomfortable questions. He was in the habit of walking around with a lit lantern during the day. When people unfailingly asked what in the world he was doing, he would respond, "I'm looking for an honest man." He had a knack for being offensive, sometimes interrupting lectures by other thinkers, such as Plato, by eating loudly.

One day, piqued by curiosity and admiration, Alexander the Great visited Diogenes, who was sunbathing in a gymnasium in Corinth. Alexander, surrounded by an entourage of bodyguards and servants, approached the great philosopher. He then made a remarkable gesture: He offered to fulfill whatever Diogenes might desire. "What is it that you'd like?" he asked. "Stand out of my light," Diogenes replied. Stunned, Alexander walked away and, according to Plutarch, said to his followers, "If I were not Alexander, I would be Diogenes."

Fund an academic you admire, and then stand out of their light. But do fund them first. Living in a barrel isn't well tolerated these days.

CHOOSING OUR HEROES, OUR LEADERS, AND OUR FRIENDS

Have you noticed how some prophets and people in power seem proud of looking like a villain in Gotham City? How being accused of sexual crimes sometimes seems to count in favor of the powerful, instead of against?[40] It's partly a matter of style, an aesthetic that reminds us of the bad boy, but aesthetics and ethics are closely intertwined. Aesthetics of dominance are becoming more prevalent.[41]

Hypothesizing why the villain is being perceived as cool is too deep a rabbit hole to follow in this book, but I suspect it is related to a misunderstanding about what it means to be a good person. The virtuous person is not the obedient one who submits to authorities and follows the rules set by them. The good person is not a pushover.

All too often virtue hinges on disobedience to unjust rules. The virtuous person is the one who is brave enough to dissent when others are following dishonorable leaders. It is Rosa Parks quietly but firmly refusing to give up her seat; it is Gandhi defying the British Empire by marching to the Arabian Sea and grabbing a fistful of salt at Dandi beach; it is Socrates "corrupting" the youth of Athens by asking dangerous questions.

Both heroes and villains tend to be outsiders and rebels. Neither bow to the normal curve. But whereas villains seek to unleash their will to power on others for their own benefit, heroes use it to protect others. Villains have a price; heroes don't. While villains pursue dominance and wealth, heroes pursue truth, beauty, and virtue. Not everyone needs to be a hero, but it would help if we got better at distinguishing the villains from the heroes. Heroes provide an ideal to strive toward, even if we never fully embody it. Tolstoy argued that it's only the person who feels her own imperfections who strives for goodness.[42]

The primatologist Frans de Waal was responsible for popularizing the term "alpha male." He came to regret it because it led to the misunderstanding that alpha male primates are bullies. They aren't. Alpha males tend to use their strength to protect the weaker members of the community. And the alpha male is sometimes not the strongest one, but the one who has tighter alliances, friends who trust him and who will support him against others. There are also alpha females. Good leaders tend to be empathic and measured.[43]

There is an alarming ratio of people recognized by *Forbes* in its 30 Under 30 list who are now convicted felons.[44] There's Sam Bankman-Fried and his colleague at FTX Caroline Ellison, both part of the *Forbes* list, both currently serving time in prison for fraud.

Charlie Javice founded a start-up called Frank that was supposed to be "Amazon for higher education," helping students navigate the process of financial aid (how that relates to Amazon eludes me). JPMorgan Chase acquired the company for $175 million in 2021. Javice said it had helped more than 5 million students. Two years later, she was charged by the Justice Department for "falsely and dramatically" inflating that number; her company had only about 300,000 clients. Javice had hired a data scientist to fabricate data for millions of nonexistent students.

Martin "Pharma Bro" Shkreli was sentenced to seven years in prison in 2018 for securities fraud. And while Elizabeth Holmes didn't make the official list, she was at the Forbes Under 30 Summit. Holmes, the daughter of a former vice president at Enron (the energy company that went bankrupt after a fraud scandal), founded a company, Theranos, and claimed it had revolutionized blood testing by developing methods that needed only a finger prick. But the machines were a sham. Holmes is serving eleven years in prison for defrauding investors, doctors, and patients.

Trevor Milton, part of the Forbes 12 Under 40 and a billionaire, founded an electric truck company called Nikola Motor. He made a video demo showing a moving truck on a flat surface. But the truck was moving thanks to its wheels and gravity; the camera had been tilted to mask that the nonfunctioning truck was rolling downhill. In 2022, he was found guilty of securities and wire fraud, and was sentenced to four years in prison, a $1 million fine, and $168 million in restitution.

I remember watching a documentary about Mexico, years ago, in which very young underprivileged children were being asked what they wanted to be when they grew up. A disquieting number of them answered they wanted to be narcos. Narcos have good cars; they are shown respect; they have power; they set up schools and sports facilities where the government has not. That they live a life of violence is less damning when violence is all around.

If our idea of what success looks like is pushing bright young people toward crime that lands them in jail or dead, perhaps we are chasing after the wrong values. The pressure to succeed early in life and in the currency of wealth misses the most important ingredients of a good life.

Sam Bankman-Fried believed there wasn't much expected value to be had after the age of forty; whatever important thing he was to do in life needed to happen before then.[45] In contrast, in *Laws,* one of Plato's last dialogues, the philosopher argues that the guardians of laws must be at least fifty years old. But Bankman-Fried didn't want to hire any grown-ups. "We tried having some grown-ups, but they didn't do anything," he said. "All they did was worry."[46] Except there were excellent reasons to worry. After it all came crashing down, Bankman-Fried asked Nate Silver, in what sounds to me like a lament, "If you go to like, May 2022—who would you have defined as the adults in the room? Who were supposed to keep me from being isolated and reality-check me and keep me honest?"[47]

The same values that are turning young minds into criminals and leaders into villains are driving our predictive machinery. It's the same will to power without wisdom driving the creation and collection of ever more sensitive data to feed predictive machines that are inching us closer to authoritarianism.

It's not easy to argue against cynicism; it never has been. It's astonishing not only that we can understand Plato, thousands of years later, but that his work is still so relevant. The philosopher Alfred North Whitehead's famous remark that philosophy is but a series of footnotes to Plato seems as true now as it was in 1929. Plato's dialogues tend to include a character who believes that power is all that matters. Thrasymachus in the *Republic* claims that justice is whatever benefits the stronger party. Glaucon puts the argument to Socrates that people act justly only when they fear consequences. Callicles in the *Gorgias* argues that conventional morality is no more than a trick invented by the weak to constrain the strong, thereby foreshadowing Nietzsche's philosophy.

Some of my students are skeptical of ethics. Against what most philosophers think, many of them have been taught in school that moral claims like "killing is wrong" are opinions, not facts.[48] Every year I have at least one young Callicles in class.

"Would you like for your child to be a psychopath?" I ask them. A nervous smile. "No." "Why not? Any good parent would like to spare their child unnecessary suffering. If your child were a psychopath, they wouldn't feel shame or guilt, they would feel much less fear for themselves, and they'd enjoy a competitive advantage with respect to others, because their lack of scruples would allow them to seize opportunities that others would balk at. Still no?" "No."

After exploring different possibilities, we tend to conclude something along the following lines: The psychopath is an appropriate object of pity, no matter how wealthy, powerful, or popular. Their life is bound to be vacuous, because they are missing out on what makes life most worth living—caring deeply about others. The psychopath cannot even imagine the depth of intimacy that is available to those willing to put another person's well-being before their own.

Instead of spending my afternoons with the messianic effective altruists while I lived in New York City, I sought the company of writers. I met Jorge Volpi, who had unknowingly inspired me to study in Salamanca, and we became good friends, walking the streets of Manhattan talking books. I signed up for a writing class at NYU taught by Antonio Muñoz Molina, who pointed me toward E. B. White and taught me to listen more closely to how people talk. I joined André Aciman's Writers' Institute, where I made lifetime friends who have been early readers for this book.

What made me want to be friends with these writers is that they all cared about beauty, truth, and virtue. I'd go into the jungle with any one of them.

DEFYING THE ODDS

*On using creativity, humor, and
resilience to flip the script*

The people we admire the most, we admire not because they followed the script they were given, but because they turned it into something else. Some people try to follow the text to the letter. Others edit between the lines and on the margins. And some others take their lot, soak it in water, turn it into pulp, and create a blank page anew. No script is more exciting than a blank page that can become anything.

Harriet Tubman's fate was to be born into slavery and die enslaved. But that was an ending she wasn't having. When she was a teenager, an overseer threw a heavy weight at another enslaved person but hit her instead. The head injury caused lifelong seizures, headaches, and vivid dreams. After escaping slavery herself in 1849, she led more than a dozen missions to free some seventy people. She carried a pistol both for protection and to prevent scared fugitives from turning back (which would have endangered the whole group). She cleverly used disguises and deception to escape detection, and developed an intricate network of safe houses. She "never lost a passenger" during her rescue missions.

Tubman was the first woman military leader in the United States. During the Civil War, she led armed raids, including the Combahee River Raid, that freed hundreds of enslaved people. She later became an advocate for women's suffrage.

ALLOWING TALENT TO THRIVE

Tubman dedicated her life to what matters. Few endeavors are as important as freeing people and defending their right to participate in their society. Her life was remarkable. But, from the point of view of an ambitious society, her genius was wasted. Had Tubman been born into a free society with universal suffrage, she could have put her extraordinary intellect, tenacity, and creativity to solving other problems humanity faces.

She could've developed a better electric car than the alternatives, and perhaps Ford would've lost that race, and today we'd be in a better place regarding climate change. She could've become a brilliant political theorist, helping us improve democracy. She could've invented something that today is still unimaginable to us. Instead, she had to spend her life, for her benefit and ours, fighting for freedoms that should be a given and could've been granted to her at birth.

Society creates its own risks and problems when it makes it more difficult than necessary for individuals to defy the odds. Life is hard enough, and humanity faces enough challenges without adding injustice to the mix.

When we deprive people of opportunities, we are dampening our collective ability to build a better world. The underprivileged child who is afraid to go to school because her clothes smell, the one we just relegated to the back of our priorities because her test results are not impressive, or because she's a refugee, might hold the answer to the problem that will threaten your life in two decades. We have reason to do whatever we can to allow her and others like her to thrive—not to mention that it's what's right.

I remember watching a documentary, years ago, that showed that when an ant column encounters an obstacle, it's the ants at the more distant fringes of the group, away from the main trail, who are more likely to be the first to find a viable new path. Through pheromone signaling, these successful alternative routes are then communicated back to the rest of the colony, allowing the entire column to adapt its path around the obstacle. We, too, depend on those on the fringes to find solutions when the mainstream hits a wall.

When previously accepted scientific theories are proven wrong, it is often the theories that were at the margin of orthodoxy that become the new paradigm. Before the 1770s, phlogiston theory explained combustion by proposing that flammable materials contained an element, phlogiston, that was akin to fire and was released during burning. Antoine Lavoisier's theory was the direct opposite of phlogiston: He thought that combustion involved combining with something from the air, rather than releasing a substance; it turned out to be oxygen. Similarly, the previously ridiculed geological theory of continental drift ended up replacing the dominant paradigm of continents fixed in their position. When we hit a wall on the main road, it's the back roads that provide a way forward.

Katalin Karikó spent her entire career studying messenger RNA, or mRNA, the genetic code that carries DNA instructions to each cell. She was convinced that it could be used to communicate to cells how to make their own medicines, including vaccines.

She climbed a steep mountain, from humble beginnings in Communist Hungary to decades of precarious contracts at American universities. She went from lab to lab, depending on senior scientists, with her unorthodox ideas mostly getting ignored and unfunded, on occasion suffering the all-too-common experience of academic bullying and harassment.[1]

Academic harassment comes in many forms: from bosses who hold a letter of recommendation ransom in exchange for being allowed to steal their mentees' work, to those who humiliate or intimidate their subordinates, all the way to rape.

Privileged academics conveniently fool themselves into thinking that harassment is an inconsequential problem, but it is a major obstacle in our search for truth. You'd be surprised at just how many careers that could've been brilliant end before they take off on account of harassment. I've known more women than I can count either who leave academia or whose careers plunge into obscurity in the throes of PTSD after suffering serious abuse, first at the hands of a harasser, followed by the complicity of universities, which tend to side with the wealthier parties (usually older men) who are more likely to be able to sue them. Harassment is an effective way for bullies to get rid of competition.

Many women end up craving marriage as a way of being protected by a husband, retreating into the private sphere of the family because the public sphere is too dangerous a place to be a woman. The harassers drive out talent we can't afford to lose, and they are more likely to be fraudsters too—the kind of person willing to do *whatever* it takes to advance their career. Bullies often have much to hide. When you are an excellent academic, you don't need to be a bully. In my experience, harassers are also more likely to plagiarize others' work, yet another great problem in academia. Often, the victims of plagiarism can prove it's their work being published as someone else's, but if that someone else is too powerful, it can seem too risky to take them on.

Robert Suhadolnik invited Karikó to work with him at Temple University—a much-needed opportunity for her. But when, three years later, she accepted a job at Johns Hopkins University, he told her that if she left, he would have her deported; her visa depended on him. He then reported her to the immigration authorities, and Johns Hopkins withdrew its offer.

As Karikó puts it, "Suhadolnik felt entitled to me. . . . [He] believed I was working for *him*. But I wasn't. I was working to solve scientific problems." Suhadolnik didn't provide a recommendation letter. Without a letter from your supervisor, you don't go anywhere else in academia, which is one reason why abuses of power are so common. Even worse, Suhadolnik spoke ill of her, making it nearly impossible for Karikó to get other positions. Eventually, and somewhat miraculously, she got a job at the University of Pennsylvania, but it never led to tenure. At one point, the university made her choose: abandon her mRNA research or get demoted. She chose demotion, with the pay cut that came with it.

Karikó's life was made hard because she was a woman and a foreigner, she didn't like to play political games, she valued meticulousness over churning out papers as fast as possible, and she had the audacity to have strong and unorthodox views. Those in the mainstream got the grants and the permanent positions while she toiled away. She was punished for being an ant at the fringes of the group.

But she was spot-on in her research pursuits. She was right to choose demotion. Her work was the basis upon which the COVID-19 mRNA vaccine was designed. Perhaps if we valued the outsider more, the outlier, the dissenter, we would better support people like Karikó, and we'd get to their lifesaving solutions sooner.

She was awarded the Nobel Prize in Medicine in 2023, after a lifetime of being belittled and ignored. Too little too late, if you ask me. We were close to missing out on her work; Karikó's formidable persistence saved us, but how many fringe ants are not as strong and resolute, how many don't choose demotion, how much talent and insights and advancements are we missing out on?

Defying the odds is at the heart of what it is to be human. It's what it means to have agency, to help build the world we inhabit. Our greatest heroes are those who achieve what seemed impossible: Abraham Lincoln, Mahatma Gandhi, Marie Curie, Helen Keller, Rosa Parks, Nelson Mandela, and beyond. They all succeeded wildly beyond expectations.

Every schoolteacher knows children who have achieved more than what was dealt in their cards. It's in our interest to have a society that allows and encourages defying the odds. Yet the more we use predictions to streamline people's lives, the more we narrow human agency, which will in turn expose us to uncharted risks. If we decide we know what someone's future will be before it arrives, and treat her accordingly, we rob her of a future that could've been different. Part of what it means to treat a person with respect is to acknowledge their agency, their ability to surprise us, to change themselves and their circumstances.

The extensive use of predictive analytics robs us of the opportunity to have an open future. By trying to eliminate uncertainty, we are also eliminating precisely the unsuspected talent that has a better chance of finding novel paths.

I once met someone who told me that whenever he applied for a job the standard way in a tech company, he got rejected with no interview; meanwhile, he regularly received offers for high-paying jobs by the same companies from people who knew him. Algorithms are not designed to identify excellence. Even people who have unusual hobbies or who went to offbeat schools can be excluded for being different.[2] When we use predictive algorithms to select people, we make the ant column narrower, pushing toward the middle those who could've been brave enough to explore the edges.

Every business leader knows that the greatest asset in a company is talent. Society's greatest resource is its citizens. If we bind people with the invisible chains of prediction, if we suffer from such hubris that we think we can know who holds promise, we are shackling our own feet.

It's not only that we are being very arrogant if we think we can afford to waste talent that could help us solve the world's most pressing challenges; it's everything else that we're missing out on, from art to humor and everything in between. Every act of creativity is a defiance of the normal curve.

VALUING CREATIVITY AND HUMOR

Algorithms rehash history. By relying too much on prediction, we are excluding the misfits and the outliers, and missing out on the extraordinary. Algorithms are better at selecting what has worked before. That's why AI, for all its disruption, is essentially conservative. That's why it tends to perpetuate sexism and racism. It's the opposite of what creativity is all about. And creativity is essential to society.

More and more, we are allowing algorithms to select our cultural products, including TV shows. Netflix was an early adopter of predictive analytics. In 2012, for example, it commissioned *House of Cards* partly based on predictive analytics. And it worked.[3] Hundreds of engineers at the company try to forecast what you want to watch. In theory, Netflix's personalized algorithm shows people what they want to see.[4] That sounds like a good thing. But is it always good? Aren't we missing something important in this model?

Seinfeld didn't start off as a successful sitcom. It decidedly *wasn't* what people wanted to watch. Test audiences thought the show was weak.

NBC's executive Warren Littlefield said that "the research report on *Seinfeld* was disastrous." "No segment of the audience was eager to watch the show again," it read.[5]

Luckily for Jerry Seinfeld and Larry David, and for fans, the NBC executive Rick Ludwin became a champion of the show, despite other people's predictions. Every film, every show, and every book needs a champion.

The network used to give Jerry Seinfeld notes on how to make the show better, but the recommendations fit generic past sitcoms. Seinfeld would listen, nod, and then say, "Okay, we're not going to do that." "You almost knew that you were on the right track when people at the network didn't like it," Seinfeld said. And his leverage was that he was always ready to walk away and go back to being a stand-up comedian.

Seinfeld attributes the unique qualities of the show to it being written by people who had never made a sitcom before. They weren't following rules, using the cookie-cutter mold, or repeating what others had done.

In one episode, three friends wait around for a table at a Chinese restaurant, which they never get. That's it. "Nothing happens!" said Littlefield. "Am I missing pages? I didn't get it." The network hated the episode, but the comedians prevailed. It is *brilliant.*

Four years later, the show was still quite obscure, attracting a narrow but loyal audience. It was surviving, though. To NBC's credit, they gave Seinfeld and David more freedom. And that's when the show went from quirky to outstanding.

In one famous episode, the four friends hold a contest over who can go the longest time without masturbating (though they never mention the word). NBC was nervous about airing it, but then again, the department of broadcast standards said that the *Seinfeld* audience had a history of going places that no one had ever gone before, so they rolled with it. It was a smash hit, winning them an Emmy Award for Outstanding Writing in a Comedy Series.

The fights with the network over the Chinese restaurant episode resulted in yet another hilarious one in which Jerry and George pitch a show about nothing to NBC. "Who says you have to have a story?" says George in that episode.

An algorithm would have never selected *Seinfeld,* but it went on to be one of the most successful shows ever produced. As Preston Beckman, in charge of NBC's research department at the time, said, "The show was different. Nobody had seen anything like it."[6] Predictive analytics working with historical data don't do new. There are no databases about the future.

Part of what is brilliant about *Seinfeld* is that it changed the audience. From being something that people didn't want to watch, it became people's favorite show. The comedy itself changed the audience's sensibilities and sense of humor. It was an acquired taste. That's the magic of creativity, good art, and good comedy: It makes us look at the world anew. Culture educates us; it affects us, changes us. People started identifying "*Seinfeld* moments" in their own lives, learning to take pleasure in absurdity that might've previously gone unnoticed, or registered only as a nuisance. Whenever I encounter the surreal in my life, and these days it happens all too often, I feel grateful for this education.

When I saw my doctoral student riding a scooter without a helmet weeks before her thesis defense, I gave her the helmet spiel about it being proof that the brain isn't even smart enough to protect the head it's in. I had invested three years in that brain! And how many times have I thought about the episode in which Seinfeld makes a booking for a car rental and there's no car waiting for him:

"I don't understand, I have a reservation. Do you have my reservation?" "Yes, but unfortunately we ran out of cars," says the woman over the counter. "But the reservation keeps the car here; that's why you have the reservation," says Seinfeld. "I know why we have reservations." "I don't think you do; if you did, I'd have a car. See, you know how to *take* a reservation, you just don't know how to *hold* the reservation, and that's really the most important part of the reservation."

Comedy plays many roles, from helping us cope with the thorns of life, to challenging power and taboos, aiding us in saying the unsayable, pointing out the absurdity of our practices, and binding us to one another in affectionate tease.

In medieval times, it was the court jester who had the privilege of speaking truths that others would've gotten punished for. If you ever find yourself in a monarch's court and are interested in the truth, don't look to the astrologer or the prophet; look to the jester.

Comedy unlocks creativity; when laughter breaks out, other shy creatures poke their noses in between the cracks. Laughter sometimes echoes in the halls of funeral homes, making grief more bearable, as loved ones of someone who lived a long and good life remember its funniest parts. Maybe I'm so attracted to comedy because philosophers strike me as comedians without a sense of humor and comedians as funny philosophers.

The London study of John Cleese, co-founder of the comedy troupe Monty Python, is decorated with books and images of lemurs, a combination that expresses well his public personality, a blend between silly and reflective.

I ask him whether he thinks AI can predict what will be funny. Cleese shakes his head. "Sense of humor is too subtle and context sensitive," he says. When I read him some jokes about philosophers created by a chatbot, Cleese is unmoved. They aren't funny. "I think it mimics rather bad comedians. As with anything, there's a lot of very bad comedians out there, and very few exceptional ones," he says.

"I've seen people argue about which joke is funnier, or how to make a joke funnier. And in the end, they always get down to matters of logic, to try to prove their point. And logic has awfully little to do with humor. Some of the funniest jokes make no sense at all.

"One joke I love is in a Marx brothers film. Groucho Marx is playing a doctor in a hotel. Somebody faints in the lobby. He races forward, kneels by this guy who has fainted, takes his wrist, looks at his watch, and says, 'Either this man is dead or my watch has stopped.' It's my favorite joke," he says as we both laugh.

Humor is elusive. "There's a scene in *Fawlty Towers* in which I'm beating my car. The first time we filmed that, I went off and found a branch, broke it off, and came back, and we rolled the cameras. I hit the car with the branch, and it wasn't funny at all; the branch was too rigid. And then I went and got a floppier branch, and that still wasn't funny; it was too floppy. Then I got the right branch, with the right degree of floppiness, and suddenly it was hilarious."

I ask Cleese about a sketch, set in an Oxbridge senior common room, in which he appears with Jonathan Miller pretending to be philosophers discussing language. The mannerisms were so spot-on that it had me guffawing on the train on my way to see him.

"I saw that sketch performed in Cambridge by Jonathan Miller and Alan Bennett when I was an undergraduate in 1962, and then that show came to the West End and revolutionized English social history. It took satire to the next level; suddenly you had people like Peter Cook impersonating the prime minister. People were shocked."

It made me think that an AI impersonating a politician would strike me as likely not very funny, because there is no person there defying power. Humor is partly about irreverence.

"Both my father and dear Captain Lancaster, who taught me Latin, thought that *Three Men in a Boat* was the funniest book in the English language. It was frowned upon by the upper classes because people laughed too much," says Cleese. "One night, when I was onstage in 1963, there were some smartly dressed people in black tie in the audience, upper-middle-class. There was an extraordinarily funny joke, and the boy in the family totally lost it; it was wonderful to see. And his father leaned across and he said, 'We don't do that in our family; we don't laugh like that.'

"One of the things humor does is that it destroys hierarchies. Once you start laughing together, hierarchy collapses. And people who are authoritarian mistrust it for that reason, because they can't stay self-important in a humorous atmosphere." I'm reminded of Mark Twain's dictum "Against the assault of laughter nothing can stand." Milan Kundera warns in *The Joke* that authoritarianism has no sense of humor. Have you felt that at airports?

"When you come up with a joke," says Cleese, "you are predicting that it will make people laugh because it makes *you* laugh. It takes empathy, experience, and understanding of context."

The best jokes of the best comedians are polished material. Cleese tells me he rewrote the film *A Fish Called Wanda* thirteen times. Even if you grant that it might take both the human being and the AI one hundred iterations to come up with the right joke, the difference is that the human being knows when they have struck gold. For the AI, it's just another iteration, because AI can't laugh.

The brilliant comedian Hannah Gadsby explains in her shows and in her book that one comedic technique is to make the audience increasingly uncomfortable before offering a release. Cleese agrees. "One of the things that you need in a joke is to build up an energy, and one of the ways you can do that best is by going into areas where the audience is already anxious." It's a very fine line, because if you go slightly too far, it's not going to be funny at all. Someone who is intentionally making you laugh is understanding something about you.

I have heard Cleese talk about the affectionate side of humor before. Joe Toplyn, a New York comedy writer, says that we are facing an epidemic of loneliness, and that if an AI companion learns humor, it can help to combat it.[7] That seems bizarre to me; the reason why humor can work as affection is that it involves someone *caring* enough to tease you.

There is emotional resonance in sharing a laugh together. Our sense of humor is a dialogue, and when we delegate the conversation to AI, we lose the personal relationship with another human being. Democracy itself is a conversation. If we give up our seat at that table and give it to an AI, we will lose our voice.

Comedy is not inconsequential for democracy. Comedy sets the tone for freedom. It gives us courage. A documentary about comedy clubs in war-torn Ukraine reminds me that humor is not a luxury; it's indispensable. "When I went to Sarajevo to a film festival," Cleese says, "people were talking about the early '90s, when Yugoslavia was falling apart. Sarajevo is down in a valley, and the Serbs were up in the hills throwing shells down, and telescopic snipers were shooting people crossing the street. And they told me that they had found an underground car park, had turned it into a cinema, and they would show comedies, including Monty Python, and they would laugh. When they came out, an hour and a half later, they felt better. Nothing had improved. But they were better able to deal with their situation."

Comedy can be an ally in our challenge to defy the odds.[8]

—

It might not be possible or desirable to offer newborns a blank page. Every one of us is born in a particular country, at a moment in history, in a family, with certain genes. Part of what we receive in our script is wisdom from those who have come before us. But we can use a pencil, instead of indelible ink. We can try leaving one blank page for every written one. We can space the lines and have wide margins to allow as much latitude as possible for edits. We'd save creatives the time and effort it takes to dissolve a script and start from scratch. We can also include a joke or two.

Allowing defiance is the right way to treat human beings, who are ends in themselves, with their own idiosyncratic views and preferences. Every life is worth living, every life a work of art to be created much more than discovered.

FREE WILL

Statistics changed how we view ourselves. But what statistics teaches us depends on how we interpret the numbers, which never speak for themselves. Data is silent about free will; it's only when we explain the data through the lens of our philosophical views that a picture arises.

In the 1830s, the most common interpretation of probability laws implied that criminals were not authors of their actions. By the 1930s, it was thought that it was precisely probability that made room for free will, what liberated us from the deterministic world of Newton. That is, probability thinking in the 1830s pushed society toward determinism, and a hundred years later it pulled it away. The dominance of predictive algorithms is pushing us back into more deterministic views of people.

The ways in which we are using predictions bring up ethical issues that lead back to one of the oldest debates in philosophy: whether we can be truly free if there is an omniscient God.[9] If God already knows all that is going to happen, that means that whatever is going to happen has already been determined; otherwise, it would be unknowable. This view, called theological fatalism, leads to the conclusion that our feeling of free will is nothing but a feeling.

What is worrying about this argument, above and beyond questions about God, is the idea that if accurate forecasts are possible, regardless of who makes them, then that which has been forecast has already been determined.

In the eighteenth century, probabilistic regularities were used to justify God. In 1710, John Arbuthnot argued that chance could not be the author of more boys being born than girls. Rather, it was God taking into account that more young men than women would be killed at sea and in war. In his demographic theology, Johann Peter Süssmilch suggested that the higher mortality rate in cities was due to sin (not bad sanitation).

A century later, statistics was used to interpret the meaning of crime. Adolphe Quetelet, the first social scientist, described criminal regularities as "a kind of budget for the scaffold, the galley, and prisons." In an 1832 letter, he wrote, "It is society that prepares the crime; the guilty person is only the instrument who executes it."[10] Quetelet saw the normal curve as implacable, and individuals as unknowing marionettes obeying its hidden command, like iron filings pulled by an invisible magnet.

The French sociologist Émile Durkheim argued that unexplained forces, external to the individual, influence society, but don't exert their pressure directly on any one individual. This mysterious force of statistics, comparable to forces like gravity, "exacts a definite number of certain kinds of actions, but not that they should be performed by this or that person." The normal curve of crime or suicides is like an epidemic striking, with some individuals resisting it better than others. "Collective tendencies have a reality of their own; they are forces as real as cosmic forces," Durkheim wrote.[11]

But here lies the tension: If each person is truly free to not commit a crime, it could happen that in a year no one does, and then the statistical law would be flouted. How can statistical forces influence groups but not individuals when groups are made up of individuals?

That contradiction led some thinkers like Cesare Lombroso, an Italian criminologist in the nineteenth century, to argue that the entire tradition of jurisprudence had to be overturned. Inspired by Charles Darwin's and Herbert Spencer's theories of evolution, he argued that if criminals are born, and not made, then punishment as retribution is immoral because criminals are not responsible.

Lombroso's proposal doesn't resolve the tension, because if every human being is entirely determined and not responsible for their actions, then so are policymakers. Determinism, taken to its extreme, undermines the purpose of arguments, as well as public policies, of any kind; *everything* is determined.

What our tour of the pitfalls of prediction in part two suggests is that the assumption that fatalism lies at the base of the ability to predict human behavior is a fallacy. The arrow of causality is pointing in the wrong direction: It is the forecast that determines the behavior.

Doesn't the way we live, and have always lived, challenge fatalism? We go through great lengths to educate our children, hoping that it will lead them to have better lives. We strive to cultivate good habits to improve our health. We praise good behavior to encourage more of it, acknowledging that people could've done worse. We punish (some) wrongdoers to disincentivize them and others from transgressing social norms, and to blame people who should've acted better. We try to structure our societies on the basis of merit.

None of those social practices that are so fundamental to our way of life would make any sense if we thought that people's destinies were sealed. Praise and blame would be inappropriate. Imagine a world without grades, fines, incentives, or punishments of any kind; a world without any attempts to change the future; a world in which people live in absolute resignation to a prophecy. It's almost unthinkable.

If the future of every company could be forecast with precision, the financial markets as we know them would instantly collapse, and with them, our economy. The whole financial market depends on us not knowing what the future holds. Though this extreme possibility of collapsing the market through perfect knowledge is impossible, do we want to go down a road that gets us closer to it?

There is an irresolvable tension between the practice of predicting human behavior and the belief in free will as part of our everyday life. A healthy degree of uncertainty about what is to come makes us want to do better, and keeps possibilities open. The desire to leave no potential data point uncollected in order to map out our future is incompatible with treating individuals as masters of their own lives.

We have to choose between treating people as machines whose future can and should be predicted, in which case anything based on merit makes no sense, and treating each other as agents, in which case making people the target of personalized predictions is inappropriate.

It would never occur to us to put a machine like a tractor in prison. If people are like tractors, then we shouldn't jail them either. If, on the other hand, people are different from machines, and we want to continue to praise and blame them, then we shouldn't treat human beings as things by predicting what they are going to do next as if they had no say in the matter.

A few months ago, I was invited to join a panel at a literary festival with an author who has written the latest iteration of a defense of determinism. The organizer asked how I would respond to the question of whether we have free will. "With a joke," I emailed back: "Do we have a choice?" He suggested that perhaps a different panel would be more suitable.

Stephen Hawking put it better: "I have noticed that even people who claim that everything is predestined and that we can do nothing to change it look before they cross the road. Maybe it's just that those who don't look don't survive to tell the tale."[12]

We're not going to get rid of fines, and grades, and policies that nudge us this way or the other. We're not going to stop educating the younger generations. So we shouldn't treat people like machines either. Having it both ways is unjust to those who lose out. They are treated by the powerful as if they were things, told that their misfortune is fate, while also being held responsible and punished for who they are or what they've done.

The more we use predictive analytics on people, the more we conceptualize human beings as nothing more than the result of their circumstances, and the more people are likely to experience themselves as devoid of agency and powerless in the face of hardship or injustice. The less we allow people opportunities to defy the odds, the more we will be guilty of condemning them, and society, to the prophecies of the powerful.

By sealing the fate of human beings through predictive algorithms, we are turning ourselves into robots. People's creativity and audacity in the face of adversity have helped to save entire nations. Think of Roosevelt and Churchill during the Second World War, who overcame unspeakable difficulties and saved the world from totalitarianism in the process. We undermine our ability to defy the odds at our peril.

Along with looking out for cars when crossing the street, we should look out for predictions too, lest democracy doesn't live to tell the tale either.

DEMOCRACY AND PROPHECY

Democracy needs an embracing of the unforeseeable to thrive. We need a healthy balance between order and chaos. Too much chaos and we risk falling into anarchy. Too much order and we risk falling into authoritarianism. If our destiny is written, there is no purpose in holding elections. It is only when an election can be won or lost, and we don't know which it will be, that democracy lives. Astrology and other forms of divination are better suited to monarchies and authoritarian regimes, where one or very few leaders rule over others, where the population's exercise of its agency is discouraged.

That astrology emerged as the Roman Republic began to collapse is no coincidence.[13] Democracy and faith in prediction are incompatible forces. The more anguishing our circumstances, the more we crave to know what the future holds. The more we consult soothsayers, the more desperate we make our situation, because by asking for predictions, we are vesting whatever power we have in fortune tellers who can seize the opportunity to enhance their power. Once the Roman Empire took over, the republic never came back.

You might agree that truth is vital to democracy, and you might even agree that the more we rely on prediction, the more we slide into authoritarian trends, but you might not be convinced that democracy is such a great system. I empathize. Democracy is frustratingly slow, inefficient, and imperfect.

And yet I fear that we are taking for granted the best invention that we've ever come up with. We tend to associate inventions with technological gadgets we can hold in our hands, but the greatest inventions of humanity are social and political.

We are forgetting the war experiences of previous generations. My great-uncle had to flee from the Francoist regime in Spain. He had to cross the Pyrenees by foot in the middle of the winter having lost a leg, gangrene eating at his stump. He had wanted to be a writer. He founded the first vanguard literary magazine in Spain and published a story in the same volume as Federico García Lorca. But after the war, he became a businessman to make ends meet. My grandfather, in turn, had been the youngest tenured professor in philosophy in Spain, but he didn't exercise his profession after the war; he was too broken.

I learned of these things years after their deaths, and years after I had studied philosophy. Knowing about where I come from made me feel part of the world in a way that I had never felt before. But the thought that I am living the life that a war denied them both is sobering.

Democracy is good for peace, stability, and prosperity; for freedom, innovation, and business. Undemocratic regimes tend to rule over businesses in intrusive and arbitrary ways. Contemporary corporate culture is one in which often only shareholders get considered when making decisions. Employees, customers, and suppliers no longer figure as prominently in the calculation—not to mention democracy. But that's new. For most of history, corporations have been aware that their privileges are conditioned on their contributing to their countries. Corporations are public institutions precisely because the government believes they will contribute to the common good.[14]

When businesses first became corporations, they had to petition the monarch or government to be granted a charter, and profit was never enough of a justification. Even during and after the Second World War, companies saw themselves as having a duty of good citizenship. Promising allegiance to stockholders only is a profitable model, but that alone doesn't make a business model justifiable. Theft and slavery are profitable too.

Many if not most corporations view themselves as stateless, with no obligation to community, either local, national, or global. One symptom of this attitude is corporations' efforts to avoid taxes.[15] Another is their brute-force approach to lobbying, spending unrivaled amounts of money to defend their immediate economic interests even when they counter democracy.[16] But businesses depend on stability and a functional rule of law to thrive. Companies are having an enormous impact on the direction of democracy, and it is in their long-term best interest to protect it, as history shows.

William Magnuson argues that Roman corporations, first created to collect taxes for the state, evolved in ways that undermined good governance. First, they threw a wrench in the gears of democratic processes by bribing legislators and sabotaging unfavorable bills. Second, they took on excessive risk, not for themselves, but for society. When the republic collapsed, largely through their fault, companies died with it, and it took a thousand years for the large corporation to be reborn.[17]

Democracy is about distributing power in a way that no one sector of society dominates the others. The heavy use of predictions with algorithms that only a handful of companies produce is skewing power in problematic ways.

In ancient Rome, when an astrologer predicted that someone would be emperor, it became difficult to sustain the stability of the republic. People's belief in fate and in the ability of seers to glimpse it made them vulnerable to losing confidence in democracy in the face of imperial predictions. Faith in prophecy could outweigh faith in the present.

That's why Rome regulated certain kinds of predictions, and so should we. The current assumption that anyone has the right to make any kind of prediction about anything or anyone through any method is harming democracy.

I have been invited by public officials around the world to advise them on known issues regarding AI ethics: privacy, bias and discrimination, the future of work, and more. The issue of the ethics of prediction has never been on the table, but that is the conversation we need to be having. Why should a tech company have the right to make a prediction about whether you live or die and sell it to insurance companies who may treat you like a dead man walking, even when that prediction might be wildly wrong?

—

When I ask Tim Harford, the *Financial Times* columnist and prolific author, if we should ban certain uses of prediction as we take a walk together near the River Cherwell, he replies, "In a liberal democracy, the presumption is tilted towards freedom; you have to have weighty reasons to ban something." Harford is profoundly knowledgeable about the theory and practice of prediction.

But what about the dangers of electoral predictions distorting democracy? "Political polls are mostly done for entertainment. The relevant question here is whether we will make better decisions if we know what other people think, or what other people think is going to happen. Part of the value of polling people is to tap into the wisdom of crowds," he says. The wisdom of crowds refers to the phenomenon whereby the aggregate judgment of many individuals produces more accurate results than individual expert opinions. It works because diverse groups bring different perspectives and errors tend to cancel each other out, while accurate insights accumulate.[18]

"But part of what it needs to work well is independence of judgment, so if you do something that influences everyone, then you are interfering with the mechanism that brings wisdom to the crowd. Voting can harness the wisdom of the crowd, but only if there is independence of judgment, and electoral predictions can interfere with that," says Tim.

Is entertainment a good enough reason to stand in the way of the wisdom of the crowd in electoral decisions? Maybe democracy is more important than entertainment. That's why we practice election silence: the banning of political campaigning, media coverage of a general election, and often the publication of polls during the election. We want to protect voters from undue influences so we can harness their insights. But in most countries the ban is only for the day of voting, or the day leading up to voting. Perhaps when it comes to electoral polls, we should extend that to give more space for people to figure out what they think for themselves.

We need a public debate about what kinds of predictions can be made about individuals, by whom, and how those predictions can be used. Profit or other prophetic interests should not stand in the way of people's autonomy, the philosophical bedrock of all rights.

Of the sources of unpredictability, the human factor is the most uncertain of them all if left to its own devices, without interferences from public predictions. Every attempt at predicting human behavior is also an attempt to control it, which in its extreme leads to authoritarianism. But human nature will not be tamed. Authoritarian regimes can crush human agency and inflict much unnecessary suffering, but humanity will topple them, sooner or later, either through acts of bravery or through the innate force of chaos latent in human unpredictability. Brute political force cannot control people entirely or forever.

It is not possible to control the human factor through science either: Predicting human behavior is bound to fail for the most extreme events, and it is also not a scientific endeavor. Those who try controlling human beings through prediction are "bound to destroy the objectivity of science, and so science itself, since these are both based upon free competition of thought; that is, upon freedom," argues Popper. The "scientific" control of human nature is both political and "suicidal." Remember the astrologer who hanged himself to fulfill his own prediction? When we endanger diversity, when we predict, control, and discipline minds, we undermine both science and democracy.[19]

For prediction to be conducive to good business practices and robust democracies, it must be limited, first, by truth and the pursuit of knowledge in contexts in which it is vital (for example, journalism), and, second, by rights, including the rights to equality of opportunity and to due process.

ROBUST COMMUNITIES

Data is silent. We make it speak. And there's always more than one way to interpret it.

In the eighteenth and nineteenth centuries, in Western countries like Britain and France, individualism and libertarianism dominated the interpretation of statistical data, argues the philosopher Ian Hacking. Their philosophers thought of states as constituted by individuals. They believed in running society through a combination of competition to incentivize innovation and productivity, on the one hand, and philanthropy for those who lost the race, on the other. In

Prussia, collectivist attitudes prevailed. Its philosophers argued that states constituted their citizens. They came up with the welfare state. Berlin offered health and unemployment insurance, as well as compensation for work accidents.[20]

Once the French idea of a statistical law entered the public debate in Europe, even the concept of law was understood differently, driven by contrasting philosophical views. The English thought about laws as given facts, whereas Germans interpreted them as a social product, as something that pertained to the culture, and not the individual.[21]

Germans like Ernst Engel saw statistical patterns not as laws but as mere regularities brought about by underlying causes. That is, Germans thought that statistical regularities were not inescapable, but something that could be modified. Engel and his colleagues, by thinking in less deterministic ways about numbers, led the way to workers' compensation, old-age pension, and a social net. They argued it was the responsibility of the state to build itself in a way that would allow individuals to become decent people and have good lives.[22]

In contrast, the English thought, disapprovingly, that beliefs in causes were a remnant of medieval metaphysical thinking, where God was the prime mover of the universe, which led them to more atomistic political models in which individuals were expected to fend for themselves. They accepted statistical regularities as explanations in themselves, without further underlying causes to search for, much less change.

We don't have to choose between individualism and collectivism. There is a middle way focusing on building robust communities in which individuals can defy their odds. A framework that is too collectivist, like Communism, will quash the individual. One that is too individualistic not only impoverishes the commons but ironically also ends up crushing the individual by not allowing for equality of opportunity. To achieve equality of opportunity, we need to respect individual autonomy while having a commons that is generous enough that any one person can stand a chance to excel. Without some kind of commons, we wouldn't have roads, firefighters, or insurance.

Insurance works only when we pool risks; if we individualize risk, we break the market. In the mid-1990s, Harvard University was running a deficit in its budget for employee benefits: It was paying more in claims than what it was receiving in premiums for its health insurance program, which had two prongs, one more generous than the other in its benefits, with Harvard paying a share of the premiums.

In an effort to correct the deficit, a new system was put in place in 1995. Instead of paying for a proportion of the premiums, the university now contributed the same fixed dollar amount toward an employee's insurance, regardless of the plan they chose. Those who chose the more generous plan now had to pay the entire $800 price difference, instead of the $280 extra they used to pay: The generous plan almost tripled in cost.[23]

The new policy made sense, at least in theory. In the abstract, it seemed fairer. With the old system, those who wanted more extensive coverage got paid more by Harvard than those who went for the cheaper option. With the new policy, those who chose to get better coverage paid the difference. Sounds fair. Except the policy ended up making *everyone's* conditions worse.

The plan with the higher coverage became too expensive, and fewer people chose it. Those who stuck with it tended to be sicker. The health costs of people who are sicker are higher, so the premiums for the more generous plans went up the following year, and Harvard found itself in the same situation as the previous year. The costs of the generous coverage kept spiraling, and within three years Harvard ceased to offer it.

No one was better off; everyone was deprived of the option of more generous coverage: The healthy (who will never be healthy forever) lost the peace of mind of having access to higher protection, the sick lost access to a much-needed resource, and Harvard ended up with less healthy and less satisfied staff.

From the point of view of economics, part of the lesson of this story is that adverse selection can kill markets.[24] Adverse selection happens when the costlier buyers, such as sicker customers, are more likely to buy insurance, and the cheaper buyers such as healthier customers self-select out of insurance coverage. But there is a deeper, more important moral lesson: Insurance markets die when they become too individualistic, when they focus too much on predicting individual fates, because insurance is about pooling risk, not individualizing it.

Harvard would've been better off either subsidizing the deficit as part of its compensation to employees or increasing premiums for everyone such that the disparity in cost wasn't large enough to disincentivize the higher coverage. The whole point of insurance is to balance out the lucky with the unlucky. If insurance doesn't allow the individual to benefit from the law of large numbers, its raison d'être ceases to be. We might as well be paying directly for our own medical bills and dispensing with the intermediary, who always takes a cut. The commission that insurers get is only justified if they do their jobs: make the policyholders better off by shielding them from bad luck.

Insurance works better when we focus on preparing as a community for misfortune in the lives of individuals, as opposed to spending effort and resources trying to predict the future—especially the future of any one person.

TO BE RESILIENT, PREPARE, DON'T PREDICT

The footage was shocking. It wasn't only the streams of water on the streets; it was the sight of what looked like tornadoes: in Valencia; not Texas, but Valencia, Spain. A photo on social media of a whole street piled with destroyed cars from one end to the other and from sidewalk to sidewalk made me suspect I was watching fake news. But it wasn't fake. The same image was there, in all the newspapers. The water was gone and the cars looked as if they had been crushed by a giant toddler god. Hundreds of people lost their lives.

I read in *El País,* the Spanish newspaper, that experts estimate that the flood in Valencia has a period of return of a thousand years. In flood analysis, the period of return is the average time interval between flood events of a specific magnitude. A thousand-year flood has a 0.1 percent chance of occurring in any given year, meaning that it is statistically estimated to occur on average once every thousand years. With climate change pressing and sea levels rising, predicting floods is about as important a job as anyone can have.

Louise Slater, professor of hydroclimatology at Oxford, has exactly that job. I ask her whether a period of return is a good way to measure or communicate the risk of floods. I worry that people might interpret it to mean that if Valencia has suffered a 1,000-year period of return flood, they are safe for the next 999 years. But the distribution of events can be clumpy; there's more than cliché to the saying that tragedies come in threes.

"That's right," said Slater, as we stroll among the huge trees in University Parks. "Another problem is that we have very little data to go on. Precise records of floods might go back fifty years, in some places, or one hundred years in other places, but we don't have data on a thousand years. Our statistical analyses usually extrapolate from data of only a few decades. We can't be sure of just how rare a major flood might be.

"And that is not even taking into account climate change. Many of our flood risk estimates assume that climatological conditions in twenty or thirty years' time will continue to resemble those of the past, but that's not necessarily the case.[25] With a warmer atmosphere, we might see floods of unprecedented magnitudes," explains Slater. It's scary to think we might be getting very important predictions wildly wrong.

"Well, we have to start somewhere, and having an imperfect projection is better than having no projection at all," says Louise. I feel uncomfortable as I think about the turkey who feels safer as Thanksgiving approaches, and about how numerical predictions can instill a sense of false security. "An even simpler approach can be to focus on risk mitigation, by using higher flood protection standards to acknowledge the uncertainty. We have to get better at acknowledging the uncertainty we face and communicating it to the public," she says.

—

Dams are a great example to think about because their fate is tightly related to climate change, and the stakes are high. Mark van der Wilk is Dutch and, although now a professor of machine learning, is an engineer by training.

"I went to Neeltje Jans, an artificial island that is part of the delta works, when I was a child. It's a system of dams designed to prevent the kinds of floods we suffered in 1953 from ever happening again. The Dutch use extreme value theory. We choose a tolerance of one failure every ten thousand years.[26] It didn't account for climate change, though, and the movable dikes will have to close more often than predicted on account of it. In contrast, the Louisiana dikes were designed to a tolerance of one failure every hundred years. That means they knowingly accept a catastrophic flood once in a generation! It's woefully inadequate."

Confronted with the challenges of climate, historians are taking a fresh look at our environmental past. Through new evidence and techniques—like analyzing isotope data from human burials, which reveals the foods people were eating, and the chemistry of clamshells, which reveals ocean temperatures—historians are teaching us just how much human affairs are influenced by climate. The Oxford historian Peter Frankopan's latest book, *The Earth Transformed,* includes dozens of examples.

The rises and falls of imperial dynasties in China are correlated with fluctuations in temperature, with colder phases bringing periods of demographic decline, conflict, and replacement of rulers. Cooler weather in England in the early modern period might've triggered an agricultural revolution, leading to new technologies, bringing about an energy transition toward fossil fuels, giving rise to European empires.

The role of climate becomes all the more relevant when human actions influence temperatures. With climate change already wreaking havoc and not showing any signs of abating, challenges to our civilizations are guaranteed.

Partly due to the unprecedented human effect on climate, and partly due to the many interacting complex elements, it is hard or impossible to predict with accuracy the effects of climate change in the long run, from the levels of water to the temperatures that will be reached and the social and economic impacts. But we can prepare.

Margaret Heffernan has a great analogy: air travel. Imagine that every time a plane took off, there was a team of engineers and data scientists trying to predict whether that day a goose will get sucked into the engine, or a part will fail, or the captain will have a health problem. It would be a very risky approach. The stakes are high; hundreds of people's lives depend on any one airplane flight, and millions of people fly every day. That's why we build planes and design flights to be robust.[27] We are prepared for a goose impacting, or a part failing, or the captain having a problem, and we don't need to spend time trying to predict these. Airplanes are built with redundancy, just as human beings have two kidneys and two lungs.

Prediction sells us the idea of leanness, but it forgets to remind us of what the price is: insecurity. It seems to work very well until it fails catastrophically. Robustness is what protects us against the unpredictable. And it's inefficient because it requires spending resources on things that you might never need. Like profit, efficiency is important, but it is not the most important value when it comes to high stakes. As Tim Harford points out, Brigham and Women's Hospital in Boston didn't forecast the Boston Marathon bombing. But they were prepared for it when it happened because they had run seventy-eight major emergency drills, covering possibilities from oil spills to train crashes.[28]

Scenario planning can be a helpful exercise precisely because scenarios are not forecasts. Thinking up and analyzing multiple possible future scenarios allows us to prepare for different outcomes and make better decisions in the present. Scenarios don't aim to accurately foretell what's around the corner. They are designed to be useful, to teach us how to think creatively and react to changing circumstances, to keep us nimble. Our muscle for invention is strengthened through navigating uncertainty.[29]

Preparation isn't only the better approach to tackling challenges like climate change and aeronautics. It's also the better approach in areas as closely associated with prediction as investment.

I ask Roger McNamee what made him such a good investor. McNamee—who is wearing a purple hoodie, jeans, and sneakers—has seen it all in tech, and is seemingly willing to talk about it all. In addition to being the author of *Zucked,* and the guitarist of the psychedelic soul and rock band Moonalice, McNamee headed the top-performing T. Rowe Price Science and Technology Fund and co-founded Integral Capital Partners, the private equity firm Silver Lake Partners, and later the venture capital firm Elevation Partners with Bono from U2. He smiles cheekily at the question. "My partners were convinced I could see what was around the corner. They thought it might be some kind of sorcery. But I wasn't trying to predict; I was preparing," he says.

"The conventional tool for analysts is to predict earnings. Some people make better predictions than others, but for all analysts and investors luck turns out to be the most important ingredient in hitting the jackpot. It seems crazy, but tech investors were not giving enough weight to the most important determinant of success for a tech company: products," says McNamee. If the main product of a tech company is unintuitive and glitchy, it's unlikely to do well in the long run. "If a tech company's product is successful, the earnings estimate from a spreadsheet will always be too low, because it's normalized," he explains. That is, it doesn't capture the extraordinary.

"T. Rowe Price started its Science and Technology Fund nineteen days before the crash of 1987. The fund was 31 percent down after a month, which was a kiss of death for a new fund," says McNamee. "Unlike me, the guys running it were rising stars, so the firm decided to redeploy them and give me the fund. I had an advantage: The prime movers were my age. The industry was new, and everyone was very young, so no one knew what was going on. They would have conferences every other week to trade intelligence; I thought if I hung around enough, eventually I would blend in, and I did. In those days, Wall Street analysts worked from behind a desk. When the best analysts were averaging twenty-nine face-to-face meetings a year with company executives, I had four hundred such meetings."

I ask Roger whether he used any numbers to come to a decision. He laughs and looks away for a second. "I used to prepare spreadsheets to show to my colleagues, but that's not how I invested; that was the marketing of it," he says, grinning. I laugh. It makes me think that big tech sometimes does something similar: They sell us the idea that they are so successful because of the data and the apps and the predictions, but that's only marketing. What made them successful is, for example, getting huge amounts of money from investors and drowning the competition through sheer economic power.

"After I had a decade of success, some people claimed that I could predict the future. But I was not trying to predict. I was concentrated on understanding which features of the present mattered more.

"After we started Elevation, Bono used to say that my superpower is that I am not motivated by money. I am always willing to walk away when others are not. That happened in 2007, when I saw Goldman Sachs was willing to put another member of the investment banking community, Bear Stearns, into bankruptcy. I realized they knew something I didn't. I had no idea what was going to happen, but often you don't need to. Another market call in my career was the dot-com crash of 2000. People have asked me how I knew what was going to happen. I didn't. What I saw was people paying billions for a company that had just added .com to its company. It made no sense, so I began to prepare for a collapse." Uniquely among tech funds, Silver Lake made profits for its investors during the dot-com crash.

Truth has a way of cutting through the bullshit, sooner or later. You're better off siding with truth in the long run. And truth is anchored to the present. In ever more uncertain times, what we need is less prediction and a firmer commitment to the present; less data analysis and more creativity; less following of prophets and more bravery standing up to them.

Predictions are the arena where fights over the future take place. Remain master of your domain, lord of the manor, king of the county, queen of the castle. Prepare for the future, and take other people's predictions as invitations to defy the odds. But how to conjure the courage to do so?

CHAPTER NINE

HOW TO THRIVE AMID UNCERTAINTY

*Curiosity, fearlessness, and philosophy
as antidotes to prophecy*

I leave the subject of death for the last class of the term in my ethics course. Why care about ethics in a finite life in which the one certainty is death? Because "until death, it's all life," as Cervantes put it.[1]

Tutorials are the essence of the Oxford system. Every week, my students write an essay that answers a question. We then spend an hour talking about their ideas in a class typically composed of two students. I teach not in a classroom but in my office. It overlooks the Bridge of Sighs, and classical music floats through the window when there is a concert at the Sheldonian Theatre.

I ask my students to forget about grades and to focus on the topic: what it means to live well. I don't get to grade them anyway (others do), and they might never again have this much time to think about the big questions in life. In a few years, they'll be too busy with jobs, mortgages, and the unending chores of adulting. For an hour, we are protected from the cold outside, the pressures of the quantified life, and prophecies of all kinds. We are even shielded from Oxford itself, from its greatness and pettiness, its beauty and oppressiveness, its luminaries and predators. For an hour, it's just philosophy.

Don't imagine an American kind of office. Mine looks more like a living room. We leave our shoes outside the room because the carpet is a light shade of beige. I sit on my yellow Poäng rocking chair, feet up on a matching stool. My two students sit in front of me, on a gray couch with two burnt-orange cushions, where I curl up and read after class. They sometimes sit cross-legged, hugging one of the cushions, their brightly colored socks contrasting against the gray couch.

Once I was offered a position at an American university. When

I visited the campus, I was shown what would be my office, a spa-cious room with beautiful trees for a view, but with furniture that looked as if it had belonged to an accountant in the 1980s. "Don't worry, we'll change the furniture," said the chair of the department, no doubt reading my face. "Could I have a couch under the win-dow?" "A . . . couch?" asked my host, eyebrows raised. "Yeah, a long couch," I said, gesturing to where it would go. "Um . . . I'm not sure if that would . . . um . . . set a good example for students," he said. I felt tempted to ask how he imagined I would use the couch, but I refrained, stifling the giggle. What I'm saying is that I don't think the kinds of conversations I have with my students would be possible in a normal classroom where I stand at the front, or even in an office in which a desk would sit between us.

Is death always a misfortune? I ask my students. About half of them think not. What does a good death look like? What does death tell us about the value of life and what exactly about life is valuable? Would you choose to be immortal if you could? What made you get up this morning? Some students talk about their friends, the boat race they are training for, their parents. Is Thomas Nagel right to argue that the value of life is so precious that it outweighs any pain? Is Epicurus correct that we have no reason to fear death because by the time it comes, we are not there anymore? Most of my students fear a death preceded by pain.

What gets you through the most painful times of your life? The answer is remarkably reliable: hope. Tell me more. Most students focus on the ideas, but some talk about their personal struggles. In their bleakest time, hope helped them survive: hope that illness would give way to health, that pain would pass, that better times would lie ahead.

A twenty-year-old has every reason to be hopeful. Most aspects of their lives will get better, and optimism is a powerful raft to hold on to. But most of my students haven't yet been through the most bitter moment of their life. If they can hope, they are not going through their bleakest night.

I expect that you won't encounter this moment for a long time, I say, but one day a doctor might tell you that your body is almost used up, that the end is near. Not everything can be fixed. What will sus-tain you then? The anxiety rises in the room as I let the silence linger.

My students look at each other, then at the floor, they fidget with their pens, they smile uncomfortably. Maybe the small pleasures in life, they say: the freshness in a gulp of water, the warmth of a sunray, the joy of good company. I nod. And what if you happen to be in pain so great that pleasure eludes you?

THE POWER OF CURIOSITY

The world can be ruthless, cruel, and unjust, but it's also infinitely interesting. What's more, in some ways, the world is at its most interesting when at its grimmest. It's curiosity that has gotten me through the most distressing times I've gone through, I tell my students.

It's in hospitals and police stations and courtrooms and the like where the makeup and the gloves come off and society reveals how things work. It's in hardship that you find out who your true friends are. It's in moments of danger and high stakes when your body and mind show you what you're capable of.

I'm not the athletic type, and I'd never climbed anything in my life. And yet, one time, I was in the middle of a peaceful protest on the Brooklyn Bridge when police arrived with buses and bouquets of white plastic handcuffs. Without hesitation, I climbed part of the bridge, from the roadway down below to the elevated walkway, one-too-many-feet above. I would've never thought I'd be the type to react that way. Hanging from the ledge, not so calmly noting my palms sweat, I remember considering whether this was a *break both your legs but live to tell the tale* kind of risk or more of a *ketchup packet going splat* one. I was not strong enough to pull myself over the ledge, and if a pair of kind strangers hadn't helped, I would've found out the answer.

Hospitals are my biggest challenge, their sickly smell of disinfectant making my knees falter. Once, in an operating room, I grabbed the sleeve of the anesthesiologist who was faffing around while cheerfully chatting to her colleagues, and pleaded with her: "Make me unconscious; now, please." She did. But if I had to do it again, I'd aspire to have the wherewithal to pay attention to the conversation of the medical staff, to find the interesting and the funny in the room. Instead, all I remember is not wanting to be there. The more you find the interesting in the grim, the less traumatic the experience.

It's not that the bright sides of life lack interest. But their other qualities—the pleasure and connection they produce—are more salient. If you're doing something pleasurable, focusing on its interest can diminish the joy. But if you're going through something unpleasant, focusing on its interest takes your attention away from the brunt of it. "What we attend to is reality," wrote William James.[2]

Grim is interesting in a bare-bones way. Grim is where you get to see the seams and the entrails. Grim is where people reveal who they are. Don't seek the grim; it'll find you. When it unfailingly appears, seize the opportunity if you can. And if you can't, don't beat yourself up about it; you can't win 'em all, and you'll likely get another chance.

The challenge is to find the interesting and the funny wherever you go. It's always there, like a game of hide-and-seek, waiting to be found, the jokes and insights reserved for those who can stand to keep their eyes wide open. The worst thing that can happen to someone aspiring to develop resilience is for everything to go well—until it doesn't. We polish ourselves when we brush against the rough.

Everything is material for a writer. And we are all authors of our lives. Good stories are sculpted from shadows as much as lights. The more of a con man and a thug Alexander the Paphlagonian from chapter one became, the more stellar Lucian's exposé. Writing, even if you never publish it, creates the necessary distance to reflect on life. The greater the injustice, the more surreal the experience, the brighter the gem that can result from it. Your demons appear unwieldy and ever present in your mind. On the page, they take a shape, become manageable, even funny. And you can transform them into anything you like.

Write life. Write for your life. Right life.

If anxiety about the future is the main driver behind our craving for prediction, curiosity is one of its most powerful antidotes. Sometimes fear will win out. But the more curiosity you can manage to conjure about the present, the less apprehension you'll feel about the future because your attention will be in the here and now.

When we discuss curiosity, I hope to be planting a seed in my students' minds that can serve as an amulet they find at the bottom of their pocket in a time of need. Will the memory drown in a sea of information that doesn't seem immediately relevant? Maybe, but you never know what your unconscious might throw as a lifeboat to you in your direst hour. Perhaps one of them, when they inevitably find themselves facing hardship, in a hospital bed or a waiting room, in a funeral parlor or a police station, will find that long-lost amulet and feel some consolation, a reason to live another day, to keep reading and writing the story.

Even at its most tedious, life is interesting. Albert Camus wrote in *The Stranger* that a person who lived for only a day could spend a hundred years in prison with enough memories to never grow bored—and that's not even counting the interesting in prison.

In a pivotal scene in the comedy show *Ted Lasso*, Ted is in a pub playing darts against Rupert, the smug ex-husband of the soccer team owner, who has challenged him, believing that Ted will be an easy mark. Ted is an American college football coach who is recruited to train an English Premier League soccer team. He is also a disarmingly optimistic character whose good cheer masks his depth. The pub fills with spectators as the stakes rise. Ted needs two triple twenties and a bull's-eye to win the game—a seemingly impossible combination.

"You know, Rupert," says Ted as he prepares to throw the first of three darts, "guys have underestimated me my entire life. And for years, I never understood why. It used to really bother me. But one day, I was driving and saw this quote by Walt Whitman painted on the wall: 'Be curious, not judgmental.' I liked that."[3] The first dart lands in the triple twenty.

"All of a sudden it hit me. All those fellas that used to belittle me, not a single one of them was curious. They thought they had everything figured out. So they judged everything, and they judged everyone. And I realized that their underestimating me . . . who I was had nothing to do with it. If they had been curious, they would've asked questions. Questions like 'Have you played a lot of darts, Ted?'" The second dart lands a second triple twenty.

"To which I would have said, 'Yes, sir, every Sunday afternoon with my father from age ten till I was sixteen when he passed away.'"

Ted throws the last dart and it hits the bull's-eye. The smiling American underdog teaches the arrogant Englishman a lesson. Kindness and white knighting win the day, and the pub cheers.

David Dwan, professor of English at Oxford, once gave a talk at Hertford College on the dark side of curiosity—the one associated with jealousy and a desire to possess and dominate. He used Vladimir Nabokov's *Lolita* to illustrate his point.

My stomach sank at the heaviness of the topic. I considered making a run for it. But the strap of my bag was trapped under the chair of the colleague next to me, the topic is fascinating, and Dwan is an engaging speaker. Dwan talked about the curiosity Humbert feels for his prey.

But Humbert *doesn't* feel curiosity for Lolita. He feels most curiosity about himself, about what it feels like for him to prey on her. Every time he gets close to seeing her, noticing her tears and her pain, he stops short, glosses over it, and moves along, looking away. He never goes there, because if he did, he probably wouldn't be capable of abusing an orphaned teenager. Humbert's curiosity for Lolita is of the shallowest kind, relating to her body and what is convenient to fuel his fantasy and desire. He never asks himself what it must be like to be her. Humbert never asks Lolita how she feels.

When Pecola, a Black girl, the most delicate and vulnerable member of society, comes into a shop in Toni Morrison's *The Bluest Eye*, the shopkeeper "does not see her, because for him there is nothing to see. . . . She looks up at him and sees the vacuum where curiosity ought to lodge. And something more. The total absence of human recognition."

Although I suspect that a lack of curiosity is a more prevalent problem, Dwan is right to point out that there are cruel kinds of curiosity. I'm reminded of the Tuskegee experiment scandal, in which clinical researchers were interested in seeing what syphilis does to the body and therefore didn't offer treatment to participants (Black men) when antibiotics became available. The right kind of curiosity is driven by benevolence; it involves caring about what it's like to be another.

—

The right kind of curiosity is also one that savors the journey. It's curiosity about the why and the how. It's not the kind of curiosity that wants to jump ahead to the end of the book or the film to know how it ends. It's the curiosity of those who seek to understand. Shakespeare sets out a summary of the plot of *Romeo and Juliet* on the first page of the play. You know from the start that the star-crossed lovers will take their lives. It's the why and the how that matter. You read on to understand and enjoy the beauty of the tale.

Curiosity is comfortable with uncertainty, because when the uncertainty is resolved, the story ends, and who wants a good story to end? Once we accept that there are some things that we cannot predict, and others that we shouldn't predict even if we can, we can start to seize the opportunities that uncertainty brings. Spaces of indeterminacy are where creativity, humor, and innovation flourish. To be successful—in life, in business, and in democracy—we need to get comfortable with the unpredictable. Resolving uncertainty as quickly as possible leads us to make bad decisions, and to surrender our power to others who end up deciding our future.

Curiosity can keep you in the present, making you less vulnerable to the allure of prophets. It can make you a better person, if you're willing to candidly ask yourself what it might be like to be the person sitting across from you, and respond benevolently. It can even make you fearless.

STANDING FIRM IN BATTLE

Would you stand firm in a battle to defend your city? As an undergraduate, I thought the passages on bravery in war in Aristotle's *Nicomachean Ethics* were irrelevant to me. I couldn't imagine myself in battle. I was also young and naive enough to think about war as something near extinct. Armed conflicts are now raging in Ukraine and Gaza, among many other corners of the world, and commercial airplanes are having to make detours to take passengers from one side of the world to another.

Even discounting literal war, existence is a struggle, and bravery a necessity for facing life's trials with integrity—from staying alive to doing what's right. It's tempting to try to play it safe by default, going for the safe career, the safe silence, the safe avoidance of confrontation. But safe bets are still bets, and not as safe as they might seem.

The illusion of safety can throw us into the fire we were aiming to avoid. The targets of bullies think they'll be safer if they comply with the aggressor's demands for silence, which only makes them more vulnerable. The rich hire bodyguards who end up plotting against them.[4] Hypochondriacs chance unintentional harm caused by doctors.[5] Germophobes poison themselves with disinfectants that weaken their immune system. Businesspeople buy the latest technology that promises a reduction in risk, only to find themselves out of pocket, dependent on a brittle technology, and subjected to the tyranny of algorithms that work for other corporations.

Courage is hard. Our instinct for self-preservation pushes us to run away when we should stay, turn a blind eye when we should look, keep quiet when we should talk, obey the rules when we should break them, and follow the market when we should lead it. And focusing on being brave just makes the task harder, much as trying not to picture a red car will inevitably bring a red car to mind. How, then, to conjure courage?

Bravery is more likely to leave us when we let our minds gallop into the future.

When my grandfather was a child, he had an accident at school. Someone shut a metal door on his finger. In an attempt to stop the bleeding and assuage the pain, he wrapped his shirt around his gory finger. When his mother saw him approaching his home with his shirt covered in blood, she had a heart attack, and died shortly thereafter. She thought he was seriously injured. She couldn't think of an alternative explanation. Had she waited for her son to be nearer, had she tempered her worst fears, perhaps she could've spared her heart from the shock of accepting as fact what was a prediction. Instead, my grandfather must've grown up feeling as if he had killed his mother.

Most of the thoughts that cause us anxiety are about the future, and most of our worst predictions will never come to pass. How many sleepless nights have you spent worrying about something that never happened?

Curiosity can make us brave. Hunger for knowledge and understanding is part of what pushes war correspondents into the trenches, and investigative journalists into the wolves' dens. When my students' self-confidence falters, I encourage them to focus on the subject matter. Research is not about them, and it's not their job to judge themselves. Their job is to do the best work they are capable of. Let other people judge it.

Courage fails us when we worry about ourselves, instead of a topic of research, the pain of others, or ideals like justice and beauty. Bravery comes along for free when you focus the lens of curiosity on the world beyond your skin.

Someone I know was recently put in the uncomfortable position of having to report serious wrongdoing. Is it worth it? people asked. Is it worth the excruciatingly long process, the danger to her own position at her institution, the dim prospects of justice? But that very question presupposes a utilitarian framework. It suggests that the right way to go about deciding something like that is to make a calculation of the costs and benefits. The more appropriate questions are whether it was the right thing to do, and whether the cause was important enough. Life is finite, and not every battle can be fought. But some battles need to be fought, whether we win or lose, because it's what's right and they matter, and there's nothing more to it.

We don't tend to be utilitarian about family relations. If someone were to ask you whether your child is worth your effort, you'd rightly be offended by the question, irrespective of the prospects of success. You do right by your child because that's what's called for, and asking whether it's worth it is inappropriate, something like a category mistake. The same goes for our duties as friends, professionals, and citizens.

Sometimes when I remind people of their duty as citizens I get asked whether it's fair to put some of the burden on the shoulders of individuals. It's not. We shouldn't have to fight for our rights for them to be respected. But the fact of the matter is that our rights are not being respected, and if you don't fight for them now, much more unfairness awaits further down the road.

To stand firm in battle, tether your mind to the present, focus on caring for people and doing what's right, and surrender to curiosity. A child reading a book at night with a flashlight under the covers, braving the anger of her parents, isn't trying to be courageous; she just wants to keep on reading. Living well has much in common with reading well, writing well, and publishing well.

BOOKS, LAST ROUND

I keep coming back to books not only because I love them and, since you've gotten this far, I assume you are partial to them too. First, books are central to prophecies because reading and writing are metaphors for deciphering the cosmos and navigating life, and books have been taken to be prophetic throughout the history of prediction. Ancient metaphors are part of who we are and how we make sense of the world.

When the Old Testament gained its canonical status, it was used as a collection of divine oracles. Ancient Jewish bibliomancy was used to investigate what God had planned for each of us. Military commanders would unroll the Torah scroll and interpret the first line their eyes landed on as instructions from God.[6] *The Book of Creation,* a work of Jewish mysticism, teaches that God made the universe through the twenty-two letters of the Hebrew alphabet. These letters have creative power; they're not only symbols but cosmic forces.

Second and most important, books have been a fundamental challenge to prophecies, being one of the most trustworthy ways of passing the torch of centuries of accumulated knowledge and wisdom. Emma Smith, professor of Shakespeare studies at Oxford, calls books "portable magic." Without books, the best that has happened to humanity would've been forgotten, as Irene Vallejo reminds us in her superb book *Papyrus.* They can speak truth even through fiction.[7] Books connect us to other people's experiences across the globe. They even allow us to talk to the dead.

Books are made to unite humanity, to help us understand one

another. They allow us to live many lives in the span of one. We can be both women and men, children and adults, and inhabit two places at once. A novel in hand gives you the chance to take a break from your life and dwell in that other one until you're ready to come back. I knew that I had found my compatriots when I read, in Marguerite Yourcenar's *Memoirs of Hadrian,* "The true birthplace is that wherein for the first time one looks intelligently upon oneself; my first homelands have been books."

Books form us. They invite us to think thoughts that might've never occurred to us. They continue a conversation stretching back to the oldest ancestors who learned how to store their experiences in words. They educate us. Tyrants fear and burn them for good reason. In the digital age, reading a paper book has become an act of defiance, of rebellion against data collection, notifications, and misinformation. When you're reading a paper book, tech can't touch you.

But if books—authors, agents, and publishers—were to follow the path of the data-driven and the prophetic, they would lose their force as valuable counterweights to prediction.

"When you read a book that has been submitted to you," I ask my literary agent, Andrew Wylie, "are you trying to predict whether it will sell well?" "No. It's always about the quality of the book. A firm instruction to anyone who works here is to not talk to me about the money. If you come to me and say 'I think this book will be a bestseller,' my response would be 'We're probably not the right agency for you.' The bestseller list is about short-term value. We're interested in enduring value."

Wylie's clients include estates like those of Albert Camus, José Saramago, and Susan Sontag, and writers like Chimamanda Adichie, Salman Rushdie, and Sally Rooney, among some fifteen hundred others. He has built a business around the books that he wants to read. Like Roger McNamee, he changed the nature of the business to fit his interests, values, and talents. (I often get asked by philosophy students whether they will fit into the discipline. That's the wrong way to think about it; you change the job to make it fit you.)

My favorite anecdote about Andrew Wylie is that he sang the

Homeric verses in the original Greek to both Muhammad Ali and Ezra Pound, the very songs that our pilgrim heard when climbing up to the Oracle of Delphi in chapter one.[8]

What makes Wylie a great custodian of books is his antithetical take on data. He isn't interested in the data, because it is irrelevant to assess the quality of a book. He knows books, and reading the words is all he needs. The numbers will follow the words.

Someone who makes a choice based on their judgment owns up to that decision and its consequences. Someone who makes a data-driven decision is shirking responsibility, deferring to the numbers. Data is replacing creativity, critical thinking, and quality in too many endeavors. The irony is that while not relying too much on data for a decision seems risky and brave, it can be a much safer approach when you're an expert, because expertise is fastened to good judgment, whether it's books, computer products, or whatever else. At best, data can represent quality by proxy.

Data-driven literary reports cite information such as similar books and how they have sold, the sales data for the author's previous book, and how many followers the author has on social media. Someone completely invested in books from a mere numbers perspective could, in theory, think it unnecessary to read the book once they have the data. Thankfully, editors tend to be interested in the words. But the more we emphasize data instead of literary quality, the more we approach Johann Bernoulli's focus on the measurement of paintings at the expense of their artistic value.

It's only when authors are encouraged to write the best book that they possibly can—not the kind of book that follows the market or caters to the data barons—that trailblazing stands a chance. It's only when committed agents and editors stand behind a book that words can prevail over numbers.

When your best client is Amazon, as is the case for books, it's hard to resist the pressure to conform to the data ideology.

"The biggest shift in the past couple of decades is where people buy their books," says Nihar Malaviya, the CEO of Penguin Random House, from his office in New York. Twenty years ago, people bought books in bookshops. Today, a significant proportion of readers get their books online, where algorithms have replaced the personal recommendations of booksellers and the aisles of bookshelves and book displays.

Covers are designed for thumbnails. Subtitles are adapted to engine searches, their keywords cramped into long, clunky sentences striving to be found among an ocean of dozens of millions of books online.

Concentrated power in the hands of platforms like Amazon offers the temptation and the opportunity to influence the market. Not even publishers could know for sure whether an online platform might've algorithmically made a book hard to find, perhaps because it was critical of tech or of a powerful politician who called in a favor. At the very least, the data collectors have an interest in preferring titles and formats that increase their revenue.

E-books are more profitable for the industry because the cost of producing them is lower than print books, and they provide reams of information about you that can be commercially exploited.

"It seems that Amazon is trying to drive out paperbacks," points out Wylie. The practice used to be for paperbacks to come out after a year of hardbacks. Those willing to pay a premium would get the hardback, which allowed them to read a book as soon as it was out, and in a beautiful and robust format. Those who could wait or wanted lighter books to carry around in a backpack would get the cheaper paperback.

Amazon is now often selling hardbacks at a cheaper price than paperbacks, and some books are not coming out in paperback at all. The incentive seems to be to buy either a hardback or an e-book. But e-books are no replacement for paperbacks.

An e-book device needs charging; it relies on an internet connection; it can easily get ruined by sand, or impacts. I once cracked one by landing on it with my knee (it wasn't intentional, but I did mildly enjoy the sound of it splitting, and didn't get a new one). Try destroying a paper book with your bare hands; it's hard work. Sure, they can succumb to fire, and water, but so do e-books.

You never own an e-book, even if you paid for it, because you don't control it. Have you had the experience of the text of an e-book you bought changing against your will because the publisher was threatened with a lawsuit or decided to delete potentially offensive content?[9] You cannot lend an e-book. It cannot keep you company.

When I was a teenager, my brother July went abroad for a year and left me his favorite hoodie, which I don't remember taking off during his absence. That's how I feel about the books that surround me. They not only bring color to my walls; they are my friends, reminding me of people, places, and ideas I love, their physical presence grounding me. They are the first and last thing I see every day. I miss them when I travel. Every now and then I find myself touching their spines. Books are objects that make life more livable.

"Sometimes you look at these bookshelves," the author Robert Caro said, pointing at his books, "and I have all these memories, all wrapped up in them."[10] An e-book is an abstract file, with no smell, touch, or physical landmarks to remember. If life is analog, an e-book is dead.

The lifelessness of an electronic reader is no impediment to its spying on you, though. Amazon and other companies are surveilling you through their e-books. They are tracking what you read, how much you read, how fast you read, what you highlight, and much more.[11] E-books are not your friends, because they don't work for you; they are informants.

We should build society in a way that gives freedom the upper hand in the face of authoritarian threats. Instead, we are building a surveillance structure that is prime for an authoritarian takeover the likes of which we've never seen before. A contemporary totalitarian regime can have much more power than previous ones. It can know where you are every second of the day, what you've searched for, how you are emotionally reacting to content, whom you are communicating with, what your hopes and fears are, and what you're planning to do next.

A striking fact that invites optimism is that, for all their efforts, tech companies have thus far been unsuccessful in making e-books dominant. Printed books are doomed, read the predictions when electronic books first appeared and for many years thereafter.[12] Convenience will win the day, tech enthusiasts said. Weightless files would turn books into relics, they hoped. And yet the mighty printed book still soars above its electronic ghost. Decades later, the tactile is winning over the convenient. Paper books are the refuge for those weary of the intrusiveness of the digital. Your paper book sends you no alerts, shares no information about you, asks nothing from you.

Decisions purportedly driven by data are the new astrological decisions; both bend toward power. The push for e-books is driven not by data but by power. It's a choice that prioritizes a particular value over others—profit. When we bow to data, we are acquiescing with the companies producing the data; it's their agenda that is getting advanced, their values that are driving the bus that we're all in. Big tech isn't following the market; they are the ones setting it. It's a bad idea to let them lead the market of ideas in the form of books. It's not enough to have fearless authors, and agents and publishers willing to publish, promote, and protect them; we also need great readers, and the kind of market in which good literature is valued appropriately.

"If Shakespeare had been able to protect his rights as carefully as Walt Disney did, then Microsoft would be a subdivision of the Shakespeare Company and everything would be better in the world," says Andrew Wylie. It's funny because it's true.

Along with the responsibility of the publishing industry as a whole to stand by truth and beauty, readers have a role to play and a chance to make a difference. Reading good literature has as great an impact on the world as on yourself. Keeping high-quality books alive matters partly because they are the medium of great philosophy.

PHILOSOPHY AS AN ANTIDOTE TO PROPHECY

Ancient Greece was obsessed with divination. Although the ancient Egyptians, Babylonians, and others were heavily invested in soothsaying, for no civilization has prophecy been as central as for the ancient Greeks.

As it sometimes happens, however, that which brings a poison can also bring its antidote. Just as jewelweed often grows near poison ivy and serves as a remedy to relieve its rash, along with the Oracle of Delphi, ancient Greece gave us philosophy.

In addition to Socrates, Plato, and Aristotle, the three brightest beacons of reason and virtue, two lesser-known philosophers strike me as especially helpful to address our prophetic challenges: Thales of Miletus and Epicurus.

Thales of Miletus

Thales of Miletus is considered the very first Greek philosopher, the first thinker who broke away from myth and divination, grounding his views in observation and deductive reasoning. Like many philosophers, Thales was known for being someone not quite of this world.

Once, while walking around looking at the stars, he failed to watch where he was going and fell into a well. An amused girl laughed at him, calling him crazy for knowing more about the heavens than about the ground beneath his feet.[13] Others made fun of him for his poverty, which they interpreted as proof of the uselessness of philosophy. Thales had had enough.

From his knowledge of meteorology, he realized that there was likely to be a heavy crop of olives. With the little money he had, early in the year and with no competition, he hired for cheap all the available olive presses in Miletus and Chios (in some tellings, he bought them). When the olive season came, he made a fortune by renting out the olive presses to people desperate to use them, lest their olives rotted before they could turn them into oil.

Thales thereby brings us the first documented creation and use of options (or futures, depending on the version of the story). But much more important, as Aristotle remarks, "Thales proved his own wisdom."[14] Philosophers, like prophets, are perfectly capable of engaging in prediction for profit. But that is not what philosophy is about. By becoming rich in this way, Thales had not added value to the world, except in the form of a lesson.

Philosophy, at its best, is an antidote to the business of prophecy. Philosophy is interested in the good life, which no amount of money or power can buy.

Epicurus

Epicurus was born six years after Plato's death, in the fourth century BCE. He was eighteen years old when Alexander the Great died. Epicureanism competed with Stoicism to dominate Greek and Roman culture and is another example of philosophy as the antithesis of and the antidote to prophecy.

Although Stoicism has become trendy for some good reasons, there is much to say in favor of Epicureanism being a better response to the challenges of modern secular life. The one point in which Stoics are perhaps superior, in my view, is that they thought of virtue as an end in itself, while the most common interpretation is that Epicurus thought about it instrumentally, as a means. However, in some passages, Epicurus argues that virtue is inseparable from pleasure, and that virtue and pleasure entail each other, which suggests a view of virtue as a good in itself. Part of the pleasant life is to act virtuously, to do the right thing for its own sake.[15]

Stoicism has become popular in technology circles and more widely partly because its defenders have understated some of its least palatable aspects. Conversely, Epicureanism has often been misinterpreted, caricatured, or ignored. Both had plenty to say about divination and prophets.

Stoics believed in divination because they thought that if gods exist, they must be benevolent, and if they are benevolent, they will find a way to talk to us, to guide us. Conversely, if divination exists, then so must gods.

Stoics thought that *pneuma,* or divine breath, permeated the world; the Pythia inhaling it was what made her see the future. That divine breath also sustained *sympatheia,* a force that wove together the otherwise disparate parts of the world: the divine realms of the cosmos with all things mundane. The greater movements of the universe created a reflection in the smaller movements on Earth. The seer was one who understood the sympathetic links between things, like the appearance of a bird crossing the sky and its meaning for the upcoming battle.

In contrast, Epicureans stood in opposition to prophets. "Divination is nonexistent," wrote Epicurus in *The Minor Epitome*. In Lucian's exposing of the con man Alexander the Paphlagonian, he tells the story of how, when an Epicurean threatened to uncover the falsity behind Glycon's oracles, Alexander commanded his followers to stone the man or face a curse. The audience almost killed the Epicurean, who was rescued in extremis. Alexander also held a public burning of Epicurus's best book.[16]

The Stoics thought that each one of us acts in line with a blueprint authored by grand designers. For them, living well is about discovering the script you have been given and embracing your destiny, fitting well with Greek myths in which the hero is the one who accepts his fate. For Epicurus, even though you are significantly constrained by the material world and external influences, there is enough leeway to be the author of your own life. Some things happen deterministically, others by chance, and others by our own hand. You might get a script on a particular kind of paper, with a number of pages, and there might already be some writing in the book, but there is still a chance for you to write your own story.

While Stoics were respectful of social hierarchy as part of fate, Epicurus was a democrat. In a garden on the outskirts of Athens, he invited women and slaves to his school—an act so scandalous that his contemporaries thought it was proof of his depravity.[17] Everyone was welcome at the Garden. Epicurus knew that anyone could understand the world and achieve happiness, irrespective of gender or profession. In his will, he left the Garden to the people in his school.

In most other things, Stoics and Epicureans had similar lifestyles, with both philosophies recommending simple lives of moderation. Epicurus has a reputation for being too focused on pleasure, but this interpretation is a misunderstanding. A more accurate portrayal is that Epicurus wished to liberate people from fears and desires so they could be free to enjoy the simple pleasures that life has to offer. The state he aimed at was one free from bodily pain and mental agitation.

Epicurus thought that we should limit our desires to the simple pleasures of life, such as enjoying good food or a good conversation with a friend. Those can be satisfied. Other desires, such as the lust for wealth or fame, are empty, in that they don't correspond to a genuine need and can therefore never be truly satisfied, since you can always be more wealthy, more famous. To fall for the allure of wealth or fame is to condemn yourself to an unquenchable thirst.

I once knew someone who suffered great anxiety and depression from the pressure of sustaining a lavish lifestyle. Had he had more temperate desires, he would've had more than enough money to last him a lifetime. Instead, he was giving himself an ulcer trying to maintain the luxury car and mansion, working so hard to earn enough money that he never had time to enjoy the car or the mansion.

Abundance, for Epicurus, lies not in what we have but in what we enjoy. Epicurus thought that people desire empty pleasures only when they're in pain; if you take away the pain, the mind can rest in tranquility without craving for more.

The new prophets, the tech barons, are not freer than you. They don't sleep better than you do. They don't have better friends or family relations. They often don't even eat better than you do.[18] Epicurus's diagnosis is that they crave so much power because they're ill at ease. No content person would behave the way they do.

Elon Musk may be the richest person on the planet, but he is far from the happiest. When he got asked what advice he'd give to those who would like to follow in his footsteps, his response was, "I'd be careful what you wish for. I'm not sure how many people would actually like to be me."[19] No one, if we listen to Epicurus. As Seneca put it, people "are choked by their own blessings."[20]

When we are free from fear and desire, we are also much less likely to fall prey to prophets. In his *Letter to Menoeceus,* Epicurus sketches an argument for why we should not fear "the most awful of evils": death. Language is misleading, he claims, for when we talk about being dead, we make it sound as if we were going to experience death. But death "is nothing to us, seeing that, when we are, death has not come, and when death comes, we are not." The poet Wisława Szymborska expressed it better: "Death always arrives by that very moment too late." Death can never catch us; as soon as it arrives, we're gone.

If you can't experience death, given that you'll be dead by then, then death cannot harm you and you have no reason to fear it. Without fear of death, we are free to enjoy the best that life has to offer.

And according to Epicurus, the most important path to happiness is having good friendships.

Thales and Epicurus developed philosophy through interacting with and reacting against those who valued profit and divination over wisdom. It's our turn to rebel against our prophets. Thales teaches us that philosophy can give you the tools to become another merchant of prediction, but it also gives you enough wisdom not to want to. Epicurus reminds us that only by freeing ourselves as much as possible from fear and desire can we enjoy life and become immune to the promises of prophets.

A prophet has nothing to offer someone who is troubled by neither fear nor desire. Divination can't tempt someone who is fully enjoying themselves. The future can wait when you are immersed in reading a good novel, playing with your nieces, having a deep conversation with a friend, or shedding tears of laughter at a comedy show. Prophets lose their influence in the face of those who are not interested in worldly power.

Like Alexander the Paphlagonian, many prophets are also bullies. They are good with psychology. They read people to manipulate them with techniques that include enticement, scaremongering, and gaslighting. Like scientists setting up mazes for their rats to go through, prophets set the stage for our rat race.

Nothing is more baffling to the prophets and the bullies than when someone stops playing their games. And there is nothing more empowering than to say, no, thanks. I'm not interested in your money or your data, in your gadgets or your threats. You cannot buy me, you cannot scare me, and you cannot bullshit me. I see you. I understand what you're doing, and I'm not going to play into it. I see your predictions for what they are: power plays. And I will write about you; what you do will go on the record, and people will see you for who you are. Whether it's your local bully at work, or a tech baron crossing a line, try opting out; it can be surprisingly liberating.

Only through curiosity, bravery, and freedom can we hope to retain control over our lives in the age of AI. If we are brave now, we won't need to be as brave tomorrow. The challenge we face is orders of magnitude larger than those faced by Thales and Epicurus. Their oracles were not automated. Divination was widespread but some-what avoidable. In his Garden, Epicurus was safe from prophets. We are making divination ubiquitous, inescapable. We are baking it into every decision made about every human being: from the time their parents start thinking about conceiving them to their medical care throughout their life, their education, their dating life, the job market, the justice system, and everything in between. Some of our most misguided leaders are trying to replace human wisdom and connection with machine oracles.

There were no cameras or microphones or digital cookies or keystroke logging or fingerprinting in ancient Greece. The fully surveilled country is not a safe one; it's a police state in which citizens are subjugated by the authority in charge and anyone else able to hack that infrastructure. Politically, the only thing worse than mass surveillance is mass surveillance at the service of an authoritarian regime automating prediction for its own benefit. Remember, the biggest promise of prediction is the accrual of power to those who wield it. And those who amass too much power end up finding themselves wanting to destroy that which is within their grasp as the ultimate way to increase their power.

The digital oracles are in control of the largest economy in the world. They are experimenting with humanity on a scale we've never seen before. It is a fight for human autonomy. But do not feel powerless, because that's what the prophets want you to feel, and you are not defenseless. The prophets' power depends on *your* data and on *your* credence, on your believing in their predictions. They depend on our cooperation. We are the children of our times. The circumstances call for us to rise to the occasion.

Philosophy has never mattered more. Don't look away, don't make it easy on the tech barons to treat us like things, don't stop reading. The power of the prophet dissolves in the face of imperturbability. Meet the trickery of prophets with the curiosity of a scientist, their intimidation with the bravery of a warrior, and their predictions with the freedom of an Epicurean.

Most people in the data economy and the data-driven rat race are not ill-willed. They are just cogs in the system, imitating what everyone else is doing, caught in the trap of the fear of missing out. But that doesn't make the system any more justifiable, and neither does it exempt us from responsibility. As Hannah Arendt shrewdly warned us with the concept of the banality of evil, some of the worst crimes that humanity has ever perpetrated have been done at the hands of bureaucrats, uncritical followers of orders, "terribly and terrifyingly normal" people.[21]

Live principledly. We tend to think that we become cynics when people around us let us down: when a friend betrays us, or when those tasked with protecting us turn out to be abusers, when our leaders show their incompetence, or our rulers reveal themselves as thugs. But I think that's self-deception.

People lose faith in humanity not when they are let down by others but when they themselves let down humanity. As long as you are able and willing to uphold fundamental ethical principles, you can remain certain that it is possible to do so, and hopeful that there are others like you out there. By living well, you will gain in integrity, satisfaction, and meaningful relationships what you lose in brute power and the illusion of safety. Instead of succumbing to the fear of missing out, take pleasure in *the joy of missing out* from the trickeries of the prophets.

When the term is over, students ask what I think they should work on over the vacations or after they graduate. Travel to places that challenge your worldview, I tell them. Talk to your parents, ask them what it's like to be them, what it was like to grow up when they did, and record their answers. One day, they won't be there to tell you about it. And read. Read the ancient philosophers and the best novels humanity has produced.

Occasionally I worry. What if I had recommended more contemporary philosophical texts that would've helped them get one more point on their exams which might make the difference for getting that job they want?

But then I remember that most of my students end up in consulting, or banking, or tech. And then I think about how many recent prime ministers studied philosophy, politics, and economics at Oxford, and I feel tempted to throw in another Dickens novel for good measure. Oxford has educated twenty-six British prime ministers, and hundreds of members of the House of Commons and the House of Lords.

It's not only British politicians. Every day I'm reminded by a loud tourist guide under the Hertford Bridge that Bill Clinton, the American president, studied here. Robert Moses, the urban planner who arguably had more of an impact than anyone else on New York City, studied politics at Oxford too. And Oxford has forged the powerful of India, Japan, Singapore, Pakistan, Jordan, Australia, among many other countries. We keep a list; we're counting them; I don't write that with pride, given that good and bad leaders have the same weight in that record.[22] The bottom line is this: I might be going into the jungle with these people in a decade or two, and so might you.

Every year I get messages from young people or their worried parents asking what they should study or where they should work to ensure themselves a good career in twenty years, as if I, or anyone else, could know. I suggest following their curiosity to wherever it takes them.

Curiosity is not trivial. Dogs have their noses. We have our curiosity. Your curiosity tells you there's something there that needs to be looked at more closely. It's a compass to pay heed to.

I grew up thinking that society had it all figured out and there was little left to do. Now I think society is clueless and what we don't know is much vaster than what we do know. So much of what we build is dysfunctional and disagreeable. Injustice is more common than justice. There is much work to be done. The book of the world could do with some heavy editing. Like the ant willing to stray from the center, we need you to follow your curiosity. Do not follow the market. Do not obey other people's predictions. Do what you love. Write the best life for yourself that you can: the kindest, the most authentic, the most beautiful. Do not get discouraged by predictions. Prophecies are there to be defied.

NOCTURNE

My mother always tells me that life is an adventure.

What is fascinating to me is how vulnerable we are to predictions given that we all know how the story ends. We know that every meeting leads to a parting, that every building is fated to collapse, that every fortune is destined to ruin, and that there is no other end to life but death. There is no mystery there. One day, our very Earth will be consumed by the fire of the Sun.

I hope that you're lucky, that you may have decades to enjoy in front of you. But for each of us this might be our last day. The difficult part is not knowing how long our ride will last. But would you like to know, even if you could? If the price to pay was getting a script, rather than a blank page on which to write?

You already know how it started and how it ends. And in the middle? Well, that's where all that matters lies. Whether it be long or short, it is guaranteed to be an adventure. Make it a beautiful one.

The future is yours to write.

EPILOGUE

Ten Lessons in Prediction

This book has focused on the perils of predictions, and how prophecy's greatest promise is power, not knowledge. But I'm not suggesting that you eschew prediction entirely, of course. Every book is finite, and you already have tech companies telling you all about the useful side of prediction, so I focused on what you don't hear every day.

We have good reason to consult our weather app every morning, consider epidemiologists' forecasts very seriously, and make judgments about which neighborhood is likely to remain a nice place to live in. We cannot help but make predictions. But we can understand better what we're doing when we make forecasts, we can be wiser about other people's predictions, and we can ground ourselves more firmly in truth. These ten heuristics on prediction might be helpful as starting points—try them out, tweak them, make them your own.

1. LIVE IN THE PRESENT

"The whole future lies in uncertainty: live immediately," wrote Seneca.[1] Life is made short by our squandering our days thinking about the future. To the best of your ability, anchor your mind and body to the present, to the book in front of you, the pleasures of the senses, the presence of your loved ones, and the qualities that make something excellent here and now.

Don't predict when you don't have to. It's hard to stop your mind from running to the future. It doesn't come naturally to us to stay in the present, and it takes some practice and discipline, but the payoff is significant, both in the currency of life satisfaction and in professional success. In the valley of the blinded by prediction, the one-eyed person is the one with the good judgment to remain closer to truth.

2. WATCH OUT FOR PROPHECIES

We live such fast-paced lives that half of the time we don't even identify predictions as such. We listen to the assertions of prophets as if they were describing the world, instead of prescribing it. The first task is to realize that someone is offering you a prophecy, and not a truth statement. The second is to ask questions.

What is this person doing with this prediction? Do you have good reasons to trust them? Are they pursuing knowledge or profit? What is the prediction pushing you to do and who will benefit from it? Is the prediction based on data? What kind of data? Who collected the data and why? What are the values that drove the data collection?

3. REMEMBER THE NATURE OF PREDICTIONS

Don't be naive about prophecies. Remember that predictions are speculative, not factual. They are wishful, in that they tend to respond to the prophet's will to power. They are speech acts used as power plays; they are closer to orders than descriptions of the world. They are limited in their vision, and they are morally significant acts. In the age of AI, beware automated decisions based on predictions posing as facts. Artificial intelligence is designed and implemented by companies with financial interests often opposed to your best interest.

4. TAKE PREDICTIONS AS INVITATIONS FOR DEFIANCE

Prophecies are the grounds on which battles over the future take place. Do not obey in advance. Prophets gain their power from people believing their predictions. Do not give power to leaders who do not deserve it. Do not give bullshitters any credence. Do not surrender your future to a prophet without putting up a fight. Whether you win or lose matters less than living well. It is your right and your nature as a human being to defy the odds.

5. PREPARE, DON'T PREDICT

Most of the time, prediction isn't necessary, and it is a distraction from the much more important task of preparing for the future. Protect your downside. Get insurance for the kinds of black swan events that could break your life. Protect yourself against major illness and disasters like fires without trying to predict them. Redundancy is a virtue as important as efficiency. Build your life and your business to be robust, to withstand the unpredictable.

Expect the unexpected. "No person has been shattered by the blows of Fortune unless they were first deceived by her favors," wrote Seneca. Know that the goddess of luck will visit you. You can never know when or how, just that she will.

Exercises like scenario planning and safety drills can remind you that the future is uncertain. They can help you stay nimble, and sharpen your abilities to respond appropriately to whatever life throws at you.

6. IF YOU HAVE TO PREDICT, PREDICT SMARTLY

If you must wander into the territory of the future, don't venture further than necessary. It's safer to predict what will happen in an hour than in a hundred years. The closer you remain to the present, the more likely you are to be right.

No prediction and no amount of data can substitute for good judgment, and there is no formula for the latter. Good judgment takes wisdom, which includes an understanding of causality, but also insight into what a good life looks like, and humility in acknowledging the limits of our knowledge.

If you work at a company depending on predictions, recognize the normalizing effect of artificial intelligence, and look to add value when the algorithms miss creative opportunities.

Get clear on what is and what isn't predictable. Are you tracking something that follows a normal curve distribution or a fat-tailed one? Is there data on the threats that might endanger your project, or might the data hide an absence of evidence on what matters most?

7. IF YOU HAVE TO PREDICT, PREDICT ETHICALLY

Predicting wisely is about predicting not only intelligently but also responsibly, ethically. Try to stick to making predictions about things and not human beings. Don't make predictions about individuals whenever possible, lest you issue a verdict in the misleading form of a prophecy. If it's necessary to make predictions about people, make them about large populations, and even then take care not to turn them into harmful self-fulfilling prophecies. Don't make predictions that contribute to injustice by treating people as things, as opposed to agents. Don't contribute to the damaging of democracy through practices like surveillance.

8. AVOID BEING THE SUBJECT OF PREDICTIONS

Being the subject of a prophecy is a dangerous thing. It amounts to losing power over your own life. Often, you won't be able to avoid it; if you go into the ER in a hospital, you will be subjected to a process of triage that involves prediction, and there is nothing you can do about it. But sometimes you can minimize interaction with prophets.

Reducing contact with predictive technology is a good start. The less you engage with surveillance systems like social media and AI, the harder it is for prophets to influence your life. Seek activities that will enhance your autonomy, like reading paper books. Writing to your political representatives asking them to protect you from the oppressiveness of prophecies is a good idea.

9. INCREASE SERENDIPITY

One way to avoid the tyranny of predictions is to increase your exposure to serendipity. The most important events in your life are likely to be the least predictable. Be open to the unforeseeable. Read widely. Talk with people vastly different from you. Allow luck to strike. Try activities that might seem out of character. Surprise yourself. Publish a quirky book, film an unusual movie, redesign an everyday object from scratch. Innovate beyond the boundaries of data. Explore the offbeat. Say hello to strangers; don't let algorithms determine whom you meet. Send messages in bottles, and if you ever receive one, consider responding. Take strolls along the beach; you never know what the tide might bring.

10. LIVE WELL

To live well is to live virtuously. Benevolent curiosity will lead the way. Courage will help you stand firm in battle. Wisdom will enable you to live in the present and will protect you from the deceits of prophets. Avoiding the mentality of a utilitarian who tries to predict and measure consequences will help you focus on what matters for the right reasons, and respect principles like human rights. Tempering desire and letting go of fear will deliver a calmer mind and true freedom. To live well is to be the author of your own life.

POSTSCRIPT

AI Ethics 101: Digital Tech's Original Sins

When I was a teenager, I saw my literature teacher, Alonso Barrera, now a playwright, walking around with a novel under his arm: *Baltasar and Blimunda,* by the Nobel Prize winner José Saramago. I immediately found a copy and read it. It was only when I finished it that I read the back cover: "Once upon a time there was a love story without love words." How had he done that?

I have attempted to copy Saramago by writing an AI ethics book (mostly) without using the term. But, unlike fiction writers who never show all of their cards, because this is a nonfiction book, I will reveal mine by ending with some words about AI ethics.

WHAT ETHICS IS NOT

Ethics is often misconstrued in the public sphere. The most common misunderstanding is that ethics is boring. The claim is usually followed by heated debates about public policy or morality that stand in direct contradiction to the original opinion. I'm lucky to teach ethics, because I never fear that my students will stay silent or feel bored in class. Ethics is fascinating because it deals with the stuff of life, questions like, What are the most important elements to be happy? What is it to live a good life? How should we organize society? What do we do about abortion, or organ donation, or poverty relief? Most people have an opinion, and usually a strong one, about such matters.

The second most common misunderstanding is that ethics is

purely a matter of opinion, and every opinion is equally valid: As if the "opinion" that killing for fun is a good thing were just as defensible as the opposite view. It's easy to say outrageous things; it's harder to live by them. A society that embraced the belief that killing for fun is desirable wouldn't last very long.

The purported moral relativist usually goes on to offer as "proof" the "fact" that different cultures have different ethical codes. But the resemblances between cultures are far more numerous than the differences. People are fairly similar creatures all around the world. Whether you are American, British, Spanish, Mexican, Indian, or Nigerian, you need a certain amount of food and water to stay healthy, you need to sleep every night, and you benefit from positive social connections. Having agency over your life is important to you. We share most physical and psychological needs and interests. And that's enough to get ethics off the ground—that, and agreeing that avoiding unnecessary suffering is a good thing, because suffering is unpleasant to the kinds of creatures we are, and people have certain claims against others, such as not being harmed unnecessarily.

WHAT AI ETHICS IS NOT

People close to the field are also responsible for plenty of misunderstandings about AI ethics. Perhaps the most important one is that AI ethics stands in opposition, somehow, to either regulation or political issues.

To develop AI ethics is not to give up on regulation—quite the contrary. Good laws are based on ethics: The greater our understanding about AI ethics, the better laws we will have, and the more we will respect those laws. Anyone who has a background in medical ethics recognizes the importance of having both ethics and good laws.[1]

Similarly, to pursue AI ethics is not to ignore politics, or set it as a secondary concern. There are many philosophers who think that there is no meaningful distinction between moral and political philosophy. Politics is a crucial piece of the puzzle in avoiding unnecessary suffering, pursuing justice, and ordering society in a way that allows citizens to have a good life.

WHAT AI ETHICS IS MISSING

Scholars and policymakers are treating most of the main concerns of AI ethics as part of a list whose members are of equal standing; I think that's incorrect. Academics often focus on the details and miss the big picture.

In *Privacy Is Power,* I argued that privacy losses, and more specifically business models based on surveillance, are at the root of most ethical issues with digital technology, and stand in opposition to liberal democracy. I stand by that argument. If we didn't collect so much personal data about people, we wouldn't be making granular predictions about them.

In *Prophecy,* I'm making a similar claim: Prediction is digital technology's other great sin. If we weren't using AI to make predictions about people, AI wouldn't have as many problems with bias and discrimination, and it wouldn't have a tendency to fabricate information, thereby being the perfect tool for misinformation. More important, if we weren't trying to predict people's lives, we wouldn't be as interested in surveilling them, and thereby controlling them.

Surveillance and prediction are partners in crime in the struggle for power. There is no democracy without citizens having agency over their own lives. In being designed to misappropriate our autonomy, surveillance and prediction are digital technology's original sins.

I first had the idea for this book years ago. I couldn't believe that the first search I did for "the ethics of prediction" (what this book is really about) turned up almost nothing. Little has changed since, other than a steady increase in the use of algorithmic predictions.

We have been using prophecy since before the Oracle of Delphi. We have waged wars, married, and bet our livelihoods on account of predictions. Every day we put our life and those of others on the line based on forecasts. It's about time we thought more carefully about the ethics of prediction. When is it appropriate to make predictions? Who is entitled to make which predictions? What are ethical methods of coming up with predictions, and what are ethical uses of prediction? I hope this book has contributed by taking the first few steps in that direction. There is much more work to be done.

APPENDIX: ANCIENT FORMS OF DIVINATION

WHAT MAKES DIVINATION POSSIBLE?

In ancient times, the book of the universe had many ways of letting itself be read. All of them were possible due to the intelligibility of the world, and the connective tissues between the larger cosmos and the smaller one around us.

Divination is a two-way conversation. When you pray, you don't get an answer. But you do when you put a question to the Delphic Pythia. Divination makes us feel closer to the gods; it makes us feel not only that the gods exist but that they pay attention to us. There is a relationship between the divine and the mundane, the divine leaving signs for us to read.

The job of the diviner and that of the doctor, the farmer, the sailor, the entrepreneur, or the writer were not dissimilar. Each must read the signs of the world to understand how what is here and now will lead to what is yet to come.

Another important idea that we first glimpse in Plato is that divination is empowered by cosmic mediators called *daimones*, who serve as messengers between gods and mortals. Even Aristotle, known for his more naturalist approach, thought that *daimones* could cause prescient dreams.[1] According to Plato, while visions often come to people when in an altered state of mind, the revelation can become meaningful only when filtered through rational decipherment, or methods of divination.

METHODS

Reading Entrails

Seers used many different kinds of props and methods to assist them in reading the signs of the world through inductive forecasting. According to Cicero, "Nearly everyone uses entrails in divination." The unlucky victims that enabled seers to read the patterns of the world in their insides were usually sheep, goats, cows, and pigs. Whether the gods changed the relevant entrails at the moment of slaughter or they nudged the inquirer to choose the right animal is unclear, but ancient Greeks and Romans thought of organs as writing tablets for gods.

Bird-Watching

Another way of reading the messages of the gods was to observe birds. Zeus's eagle and other powerful birds of prey were the most ominous. Woodpeckers were a sign of good luck for carpenters, hunters, and those on their way to feasts. What the bird did and the noises it made were meaningful too. In Aeschylus's *Agamemnon,* as the Greek leaders are on their way to Troy, two eagles devour a pregnant hare in front of them. Calchas interprets the scene as meaning that the way will be long but the Greeks will triumph.

Chance Occurrences

Any kind of chance occurrence, from coincidences to spontaneous remarks, was also material for predictions to be made. And, of course, observing the skies. An Old Babylonian list of celestial portents claims that an eclipse "in the evening watch is for plagues," and one "in the middle watch is for diminished economy."[2]

Texts

For the freelance seers and magicians, there were also texts. The Greek magical papyri (*Papyri graecae magicae*) span a period between the last century BCE and the fifth century CE and contain about six hundred spells, the recipes of individual magicians, passed on from one to another. To invoke Apollo, for example, magicians must wear a crown of laurel, like the Pythia, and white robes. The laurel became Apollo's sacred plant after the nymph Daphne, whom he loved, was turned into a tree. Using laurels was akin to turning the tree back into a nymph. Other ways of achieving divinatory powers were more complicated, involving the killing of animals and the concoction of elaborate (and disgusting) potions.

Some spells were designed to allow the magician to receive prophetic dreams. Others were intended to send dreams to other people. Think of it as ancient hacking attempts. There was no way of distinguishing reliable dreams from trick ones (sometimes sent by gods themselves, or by sorcerers). The magician was hired to send a dream to someone with the hope that they would accept the dream as divinely prophetic.

But pulling off this trick wasn't easy. The magician had to take a cat and "make it into an Osiris" (also known as murdering it) while speaking a magic formula invoking Bastet, the Egyptian cat-faced goddess.[3] The process is one instance of the more general practice of achieving control over a ghost who could do one's bidding in worlds beyond this one. I'll spare you more grisly details; suffice it to say that some spells involved human heads.

Possession

At Delphi, there was a vapor rising from a chasm in the ground beneath Apollo's temple that intoxicated those who inhaled it. The Greeks believed that when disembodied souls, *daimones*, encountered the soul of the Pythia, they intertwined, thereby possessing her and exchanging information.[4]

Public Debate

One of the most important aspects of divination in ancient Greece was the debates that it gave rise to. Interpretation by way of public debate was an essential part of the process, a method in itself. Human judgment was a crucial input in discovering the meaning of a prediction.

For debate to perform its function, oracular pronouncements had to be open to more than one interpretation. Greeks' penchant for riddles fits in well with the possibility of public debate. To be a Greek hero involved excelling at solving riddles, like Oedipus solving the Sphinx's riddle about the stages of life: *What walks on four feet in the morning, two in the afternoon, and three at night?* Ambiguity was also convenient for oracles; it made their claims hard to dispute. When multiple interpretations are possible, predictions can't be obviously wrong. Everyone had an interest in predictions being vague, from the oracle to the petitioners. Metaphors, homonyms, and double meanings were the elements that gave life to the game of forecasting.

During the Persian Wars, the Oracle of Delphi told Athens to protect itself with "wooden walls." Debate ensued. Were Athenians expected to literally build a wooden wall that could be easily burned? The official oracle interpreters had suggested that they use thornbushes to surround the Acropolis. The general Themistocles ended up convincing his countrymen that "wooden walls" was a reference to Athens's navy of wooden ships. It was a wise interpretation; Athens won.

Understanding the Past

Prediction is inseparable from understanding the past and how it relates to the present. In the *Iliad,* Calchas is described as someone who "knew all things that were, the things to come and the things past." Looking to history was yet another predictive method.

When the plague besets the Greek camp, Achilles asks Agamemnon to find a seer who can tell them why Apollo is angry and how they can assuage him. Soothsayers were crisis managers whose job was to discover what had gone wrong in order to right it. Calchas explains that the problem was brought about by Agamemnon's kidnapping of Chryseis. Her father, the priest of Apollo, offered the king gold and silver as ransom to recover his daughter. Agamemnon rejected the offer, disrespected the priest by taunting him with the image of his daughter being a sex slave, and sent him away with threats of violence. Chryseis's father prayed to Apollo for revenge, which came in the form of the plague. Only with these wrongs righted would the plague subside.

Medical diagnosis likewise used the past to understand the present and mold the future. Diogenes Laertius writes that Epimenides cured the plague in Athens by tracing its problems to the ghosts of some men who had been murdered after they took refuge in the precinct of the Semnai Theai. The cure was to sacrifice a flock of black and white sheep that were to be let loose on the Areopagus, near where the men had been murdered. Wherever the animals lay down, there they were to be sacrificed, an altar to be erected on the same spot.

Dreams

Galen, the famous physician, started his career when his father had a dream about him studying medicine in Pergamon. He claimed to have saved many lives by applying a cure revealed to him in dreams.

Interpreting the Effect of Celestial Bodies

Ancient doctors in Greece and Rome, including Galen, wrote of the effects of the planets and moon on the body and mind. The term "lunatic" refers to the belief that the moon influences health, from menstrual cycles to recurrent fevers, epilepsy, and certain forms of madness. Similarly, the ancient idea of sympathy referred to the relationship between the heavens and Earth, and how different animals, plants, and stones were favorable or adverse (antipathetic) to certain conditions. Physicians used amulets made out of favorable gems for their alleged healing properties.

Ancient doctors who believed in sympathetic magic thought that in illness critical days were expressed in patterns correlated to the moon:

after seven days from the new moon, for example.[5] Depending on the stars and the moon, there were days when diseases would take a turn for the worse, and other days on which they would improve. There were also better times than others to be bled, or to have surgery.

Pliny the Elder, in the first century CE, writes about a doctor from Marseilles who prescribed food according to the positions of the planets and made a fortune by it.[6]

We've come a long way since the ancient methods of prophesying, but not so long to have lost the main belief that sustains divination: that of an intelligible world which leaves signs that can be read to look into the future. For better and worse, our vision into what will be is much more limited than we profess. That limit, along with the audience's desire and belief that the future can be knowable, creates a power vacuum, inevitably filled by those who can spin a good story about how the future will conveniently align with their interests. Reading entrails is vastly different from making an astrology chart, which in turn is vastly different from designing a predictive algorithm. But the political function of prophecy remains the same. The promise of prediction is power, and its peril the abuse of power.

ACKNOWLEDGMENTS

Thank you:

First and foremost, to readers, who make writing worthwhile. The thoughtful notes I've received from readers around the world in the last few years have motivated me to keep on writing, rain or shine.

To my students, who are willing to embark with me on conversations about what life is all about.

To my agent, Andrew Wylie, and to Katie Cacouris and James Pullen, from the Wylie Agency, for being such superb guardians of books.

To my editor Edward Kastenmeier, from Doubleday, for betting on *Prophecy* and for improving the book with the precision of a true expert. To Oliver Munday for the stunning cover, and to the rest of the team at Doubleday, including Chris Howard-Woods for the great editing, and Ingrid Sterner for the meticulous copyediting (where would books be without good copyediting?), Sara Hayet, Laura González, and Sam Bonner for the brilliant work taking care of publicity and marketing (without whom you wouldn't know this book exists). To Louise Carron and Amelia Zalcman for legal counsel. To Managing Editor Vimi Santokhi, Production Manager Peggy Samedi, Production Editor Amy Brosey, and Text Designer Michael Collica for all the work behind the scenes. It takes a small brave army to publish a book.

To my editor Miguel Aguilar, from Debate, for encouraging me as an author and for believing in *Prophecy*, and to the rest of the team at Debate—it's a pleasure to work with you.

To the Rockefeller Foundation and its Residency Program at the Bellagio Center, where this book was partly written. I'm grateful to all the residents there, but especially to Mitch Jackson, Tara Westover, and Oscar López for their valuable feedback, encouragement, and friendship.

To Ragdale, for its Human Residency Fellowship.

To Harvard University's Edmond & Lily Safra Center for Ethics—especially to its director, Eric Beerbohm—for taking me in as a fellow.

To the principal and fellows of Hertford College, good colleagues and good friends, for always offering a refuge—especially to Pat Roche, Anette Mikes, Mark van der Wilk, Ciaran Martin, Vlad Vyazovskiy, Ian McBride, Michael Wooldridge, Aruna Nair, David Dwan, Kate Greasley, and Emma Smith, as well as to Tom Fletcher, who will always be a part of Hertford in my mind.

To the members of the Institute for Ethics in AI (to Cristina Bratu especially, for helping me get the message out there), and to the Faculty of Philosophy and the University of Oxford for the support.

To the British Academy, for a Mid-Career Fellowship, and the Leverhulme Trust, for a Major Research Fellowship (grant MRF-2023-124). Without these two grants, which temporarily freed me from most of my teaching and administrative duties, this book would not have been written.

My gratitude in advance to the future benefactor who might consider funding my research so I can continue to write books; this is my message in a bottle.

To my Spanish colleagues and friends Vicente J. Montes Gan and Carlota Taboada at the Rafael del Pino Foundation, and to Carlos Luca de Tena, Alex Roche, Irene Blázquez, Lucía Egea, and Manuel Muñiz at the Center for the Governance of Change at the IE University.

To the board members of the Proton Foundation—Andy Yen, Antonio Gambardella, Dingchao Lu, and Sir Tim Berners Lee—and everyone at Proton, who prove every day that better technology is possible.

To every person who lent me their knowledge and allowed me to interview them for the book.

To Will, Timo, Reuben, and Harrison, for providing a writing space worthy of the best of sitcoms.

To my friends, who make me more of an Epicurean by the day: Ángel Maldonado, Anil Seth, Diego Rubio, Sonia Contera, Hannah

Maslen, Silvia Milano, Daniela Torres, Carlo Cordasco, Matthew Dennis, Carina Prunkl, Teresa López de la Vieja, Luciano Espinosa, Rubén López Pulido, Rosana Triviño, Josh Shepherd, Alberto Giubilini, Andrea Rizzi, Rupert Younger, Rafael Ramírez, Rafo Mejía, Marina López-Solà, Antonella Mallozzi, and Landon Lobb, among others. To Richard Ovenden and Andre Stern. To Chris Wylie and Carole Cadwalladr. And profound thanks to K., for being all in.

Special thanks to the friends who read early drafts of parts of this book: Roger McNamee, Nigel Warburton, Jorge Volpi, Milo Phillips-Brown, Mark van der Wilk, Roberto Trotta, James Knight, Tom Harper, Sheila Miller, Annie Dycus, Jon Durbin, Bill Hood, Lisa Williams, and Karen De Luca. Nigel Warburton sat next to me in our favorite coffee shop morning after morning, writing his own book. Roger McNamee was the first person I talked to about this book and has cheered me on every step of the way; thank you.

To Marisol Gandía, Sole Vidal, Silvia Gandía, and the rest of my Valencian family.

To the kindness of strangers, who have often become friends, and have made my life better by helping out innumerable times, in small and big ways, even when they stood to gain nothing from it.

To my family, for their curiosity and warmth, for giving me a blank diary in which to write life and encouraging me to follow my own path, for reminding me that life is an adventure, and for unfailingly standing firm in battle, by my side, every time I've set out to defy the odds.

NOTES

PRELUDE

1. Rescher, *Predicting the Future*, 1.

OVERTURE: WHAT'S IN A PREDICTION?

1. Matsumi and Solove, "Prediction Society."
2. Bickel and Kim, "Verification of the Weather Channel Probability of Precipitation Forecasts."
3. Yates, *How to Expect the Unexpected*, 376–77.
4. Vincent, "Instagram Internal Research." For an excellent film on the consequences of such decisions, see Marc Silver's *Molly vs The Machines* (2026).
5. Forst, "Noumenal Power."
6. Weber, *Economy and Society*, 53.
7. Arendt, *Origins of Totalitarianism*, 457.
8. Ambady, "Perils of Pondering."
9. Rothman, "What Happened to the Car Industry's Most Famous Flop?"
10. Ferens, "Edsel."

CHAPTER ONE: PROPHETS AND POWER

1. Johnston, *Ancient Greek Divination*, 48.
2. Diodorus Siculus 16.26.1–4.
3. Johnston, *Ancient Greek Divination*.
4. Ozalp et al., "'Digital Colonization' of Highly Regulated Industries"; Kraaijeveld and Sharon, "Increasing Influence of Big Tech in Health and Medicine and the Need for a Public Health Ethics Perspective."
5. Guyer, "Big Tech's Push into Military AI Is Troubling."
6. Frenkel, "Militarization of Silicon Valley."
7. Fields, "False Prophets and Fake Prophecies in Lucian."
8. Olomi, "Baghdad."
9. Barton, *Ancient Astrology*, 56.
10. Tacitus, Annals 6.22, trans. M. Grant.

11. Augustus wasn't the only emperor to use astrology to legitimate his place at the top of the hierarchy. More than a century later, Septimius Severus, conscious that his predecessor had lasted only three months in power before being murdered, published dreams, oracles, omens, and other prophecies foretelling his success as emperor in his autobiography, and ordered they be represented in public sculptures and paintings. Barton, *Ancient Astrology*, 46.
12. F. S. Naiden, "The Use of Divination by Macedonian Kings," in *Divination and Prophecy in the Ancient Greek World*, edited by Roger D. Woodard (2022), 233.
13. Kay, "Spare Ribs of Dante's Michael Scot."
14. Suetonius, *Twelve Caesars*, Vitellius, 14.4.
15. Barton, *Ancient Astrology*, 65.
16. Thanks to Antonio Gambardella for first mentioning this term to me.
17. *On Fate and Providence*, PG 50.756–58.757, 8.
18. Theodosian Code 9.16.4.
19. Pliny, *Natural History* 7.49.160ff.
20. Lewis, "Reviewed Work: The Crime of Claudius Ptolemy by Robert R. Newton."
21. Fourny, "Élizabeth Teissier raconte ses rendez-vous avec Mitterrand et Chirac."
22. Diente, "Princess Diana 'Predicted' She Would Die in a Staged Car Accident, Says Advisor."
23. Griffiths and Culliford, "Princess Diana's Astrologer Read Her Horoscope to Her a Month Before She Died."
24. "Astrologers Are Predicting the Result of America's Election," *Economist*.
25. Plato, *Republic* 2.520d–521b.

CHAPTER TWO: CONQUERING UNCERTAINTY AND TAMING RISK

1. Heinecke, "When Marie Curie Was Almost Excluded from Winning the Nobel Prize." Other symptoms of sexism include the Swedish Academy failing to protect women from the sex offender Jean-Claude Arnault, the inexcusable gender imbalance of the laureates—out of 229 physics prizes, only 5, or 2 percent, were given to women—and the academy's habit of ignoring women co-authors in jointly awarded prizes.
2. Social media post by Timnit Gebru. www.linkedin.com/posts/timnit-gebru-7b3b407_google-deepmind-co-founder-demis-hassabis-activity-7249908935 699132416-ZLGU.
3. See the work of Sasha Luccioni on AI and the environment, as well as Kate Crawford's *Atlas of AI*, among others.
4. Shanbhogue, "Will the Bubble Burst for AI in 2025, or Will It Start to Deliver?"
5. Seth, *Being You.*
6. Vincent, *Beyond Measure*, 96–98.
7. Ibid., 99–103.
8. Quoted in Shapin, *Scientific Revolution*, 32.
9. Galileo, *Discoveries and Opinions of Galileo*, 237.
10. Hacking, *Emergence of Probability*, 41.

11. Ibid., 32.
12. Ibid., 40.
13. Ibid., 37.
14. Hacking, *Taming of Chance*, 14.
15. Ibid., 20.
16. Bernstein, *Against the Gods*, 126.
17. Chivers, *Everything Is Predictable*, 9.
18. Bernstein, *Against the Gods*, 131.
19. Chivers, *Everything Is Predictable*, 44.
20. Vincent, *Beyond Measure*, 165–69.
21. Hacking, *Taming of Chance*, 92.
22. These numbers are based on those of the Office of National Statistics. About 70,000 rapes were reported between October 2023 and September 2024, and it is estimated that 5 in 6 women who suffer rape don't report it. www.ons.gov.uk/peoplepopulationandcommunity/crimeandjustice/bulletins/crimein englandandwales/yearendingseptember2024; www.ons.gov.uk/peoplepopulationand community/crimeandjustice/bulletins/sexualoffencesinenglandandwales overview/march2020; rcew.fra1.cdn.digitaloceanspaces.com/media/documents /Rape_and_sexual_assault_statistics_sources_January_2025.pdf.
23. These numbers come from the Crown Prosecution Service statistics, as well as the House of Lords Library: lordslibrary.parliament.uk/rape-levels-of-prosecutions.
24. "Decriminalisation of Rape."
25. Ward, *Loop*, 254–55.
26. Hacking, *Taming of Chance*, 43.
27. Trotta, *Starborn*, 184–85. I highly recommend this book to anyone wanting to know more about the stars and why they matter.
28. Gillispie, *Pierre-Simon Laplace*, 65, 276.
29. Ibid., 274.
30. Vincent, *Beyond Measure*, 218–19.
31. Trotta, *Starborn*, 187.
32. Dunnington, *Carl Friedrich Gauss*.
33. Trotta, *Starborn*, 186.
34. Vincent, *Beyond Measure*, 219.
35. Schweizer, "Unraveling the True Identity of the Brain of Carl Friedrich Gauss."
36. Hacking, *Taming of Chance*, 105.
37. Porter, *Rise of Statistical Thinking*, 102–3.
38. Hacking, *Taming of Chance*, 110. As my friend Roberto Trotta pointed out to me, this point may be incorrect: The *n* measurement of the height of a single soldier will converge toward his true height, and so becomes more and more precise with more measurements. The *n* measurements of *n* different soldiers will follow the soldiers' population distribution, which does not vary with *n*. Therefore, one should be able to tell them apart. But Quetelet covered himself by stipulating that the hypothetical measurer was "little practised in measuring the human body."
39. Ibid., 144–45.
40. Galton, "Regression Towards Mediocrity in Hereditary Stature."

41. For more on the links between eugenics and the tech culture that is designing and managing AI, see Bender and Hanna, *AI Con*, chap. 2.
42. Hacking, *Taming of Chance*, 182.
43. Galton, Presidential Address, *Journal of the Anthropological Institute* 15 (1886).
44. Clayton, *Bernoulli's Fallacy*, 134.
45. Ibid., 158.
46. Hacking, *Taming of Chance*, 183.
47. Birhane, "Impossibility of Automating Ambiguity."
48. Hacking, *Taming of Chance*, 48.
49. Einav, Finkelstein, and Fisman, *Risky Business*, 72.
50. Hacking, *Emergence of Probability*, 5.
51. Graeber, *Debt*.
52. Bernstein, *Against the Gods*, 90.
53. Ibid.
54. Porter, *Trust in Numbers*, 105.
55. Dahms, "Diligent Bureaucrats and the Expulsion of Jews from West Prussia."
56. Hacking, *Taming of Chance*, 67.
57. Muller, *Tyranny of Metrics*, 40.
58. Porter, *Trust in Numbers*, 200–211.
59. Ibid., 200–201.
60. Ibid., 8.
61. Vincent, *Beyond Measure*, 70–71.
62. Chevalier, "Opening Address."
63. Hacking, *Emergence of Probability*, 18.
64. Porter, *Trust in Numbers*, 37.
65. Ibid., 43.
66. Ibid., 77.
67. Ibid., 157.
68. Ibid., 177–78.
69. Ibid., 97.
70. Rosa, *Uncontrollability of the World*.
71. Arendt, *Origins of Totalitarianism*, 318.
72. Hacking, *Taming of Chance*, 146.
73. Hacking, *Emergence of Probability*.
74. Porter, *Trust in Numbers*, 213.
75. Ibid., 37.
76. Gigerenzer and Murray, *Cognition as Intuitive Statistics*.
77. Porter, *Trust in Numbers*, 21.
78. MacIntyre, *After Virtue*, 107.
79. Vincent, *Beyond Measure*, 17.
80. McGrayne, *Theory That Would Not Die*, 14.

CHAPTER THREE: THE ULTIMATE PREDICTION MACHINE

1. Laplace, *Philosophical Essay on Probabilities*, 4.
2. Harford, "Astonishing Record—of Complete Failure."

3. Casselman, "Economists Are in the Wilderness."

4. For more on economic forecasting, see Sherden, *Fortune Sellers;* Friedman, *Fortune Tellers.*

5. For an accessible history of AI, I recommend my colleague Michael Wooldridge's *Brief History of Artificial Intelligence.*

6. Crawford, *Atlas of AI.*

7. Dan Hendrycks (@DanHendrycks), X, Sept. 9, 2024, 10:37 a.m., x.com/DanHendrycks/status/1833152719756116154. Thanks to Mark van der Wilk for sending me this example.

8. Hill, "Your Car Is Tracking You."

9. Krause, "Case Study: Amazon's AI-Driven Supply Chain"; Mollick, *Co-Intelligence,* 6, 7.

10. Obermeyer et al., "Dissecting Racial Bias in an Algorithm Used to Manage the Health of Populations."

11. Moore, "Blame AI for Your Gruelling Job Interview"; Jaser and Petrakaki, "Are You Prepared to Be Interviewed by an AI?"

12. "Use of Artificial Intelligence in Government," National Audit Office, March 2024.

13. Macintyre, "Precrime Profiling Is No Longer a Fantasy."

14. Gellman, *Dark Mirror;* Tau, *Means of Control.*

15. Angwin et al., "NSA Spying Relies on AT&T's 'Extreme Willingness to Help.'"

16. Wilson, "Private Firms Provide Software and Information to Police, Documents Show."

17. Ng, "Google Is Giving Data to Police Based on Search Keywords, Court Docs Show."

18. Biddle, "How Peter Thiel's Palantir Helped the NSA Spy on the Whole World."

19. Gerstein, "Anti-Democratic Worldview of Steve Bannon and Peter Thiel."

20. Wylie, *Mindf*ck,* chap. 1.

21. "Facebook Stops Sending Staff to Help Political Campaigns," *BBC News.*

22. Dylan Byers and Ben Collins, "Trump Hosted Zuckerberg for Undisclosed Dinner at the White House in October," NBC News, Nov. 21, 2019.

23. Barnett, "Record-Breaking 2024 US Political Ad Spend Likely to Disrupt Commercial Media Landscape."

24. "Online Political Spending in 2024," Brennan Center for Justice, Oct. 16, 2024, www.brennancenter.org.

25. "Online Ad Spending in the 2024 Election Topped $1.35 Billion," Brennan Center for Justice, Dec. 19, 2024, www.brennancenter.org.

26. "Revealed: Trump Campaign Strategy to Deter Millions of Black Americans from Voting in 2016," Channel 4 Documentary, www.youtube.com/watch?v=KIf5ELaOjOk; Véliz, *Privacy Is Power,* chap. 4.

27. Henshall, "There's an AI Lobbying Frenzy in Washington."

28. Madison Hall, "Tesla, Amazon, Microsoft, and Other Tech Giants Are Spending Millions in DC. Here's Who Is Spending the Most," *Business Insider,* May 10, 2024.

29. Schaake, *Tech Coup,* 59–60.

30. Ibid., 28–29; April Glaser, "Thousands of Contracts Highlight Quiet Ties Between Big Tech and U.S. Military," NBC News, July 8, 2020.

31. Frenkel and Krolik, "Trump Taps Palantir to Compile Data on Americans."

32. Brian Fung, "Pentagon Awards Multibillion-Dollar Cloud Contract to Amazon, Google, Microsoft, and Oracle," CNN, Dec. 8, 2022.

33. Schaake, *Tech Coup,* 28.

34. Olson, *Supremacy,* 257.

35. Rana Foroohar, "Should Tech Giants Be Treated Like Nation States?," *Financial Times,* Nov. 18, 2024.

36. Stokel-Walker, "Revealed: How the UK Tech Secretary Uses ChatGPT for Policy Advice."

37. David Folkenflik, "Over 200,000 Subscribers Flee 'Washington Post' After Bezos Blocks Harris Endorsement," NPR, Oct. 29, 2024.

38. Duffy and O'Kruk, "How the Feud Between Elon Musk and Donald Trump Exploded over 72 Hours."

39. Burns and Otterbein, "Trump Staff 'Furious' After Musk Trashes AI Project."

40. Sloan and Maloney, "Elon Musk's Feud with Trump Spurs One of His Worst Wealth Losses Ever"; Schleifer, "Elon Musk Still Holds a Bit of Leverage over President Trump."

41. Ponciano, "Google Billionaire Eric Schmidt Warns of 'National Emergency' if China Overtakes U.S. in AI Tech."

42. Daniel and Palmer, "Google's Goal."

43. Jenkins, "Google and the Search for the Future."

44. Empathy Holdings has invited other companies to a conversation on commerce, technology, and ethics via Ethical Commerce Alliance.

45. Schaake, *Tech Coup,* 130. Musk has denied this interpretation. He claimed that he didn't intend to turn off Starlink.

CHAPTER FOUR: LIKENESS TO TRUTH

1. Hsee and Hastie, "Decision and Experience."

2. Some paragraphs in this section first appeared in Carissa Véliz, "What Socrates Can Teach Us About AI," *Time,* Aug. 1, 2023.

3. Frankfurt, *On Bullshit.*

4. Vallor, *AI Mirror;* Véliz, "Why LaMDA Is Nothing Like a Person."

5. Weintraub, "Chatbots Are Accurate, Show More Empathy than Doctors in Answering Questions, Study Finds"; Ayers et al., "Comparing Physician and Artificial Intelligence Chatbot Responses to Patient Questions Posted to a Public Social Media Forum"; Kolata, "When Doctors Use a Chatbot to Improve Their Bedside Manner"; Vowels, "Are Chatbots the New Relationship Experts?"

6. Véliz, "Chatbots Shouldn't Use Emojis."

7. Hill, "Why Is ChatGPT Telling People to Email Me?"; Field, "Doctors Fear ChatGPT Is Fuelling Psychosis."

8. Klee, "People Are Losing Loved Ones to AI-Fueled Spiritual Fantasies."

9. Klee, "He Had a Mental Breakdown Talking to ChatGPT."

10. Megan Cerullo, "Air Canada Chatbot Costs Airline Discount It Wrongly Offered Customer," CBS News, Feb. 19, 2024.

11. Aristotle, *De Interpretatione* 19a25–30.

12. Anscombe, "Aristotle and the Sea Battle."

13. Chelli et al., "Hallucination Rates and Reference Accuracy of ChatGPT and Bard for Systematic Reviews."

14. Matsumi and Solove, "Prediction Society."

15. Heaven, "Hundreds of AI Tools Have Been Built to Catch Covid."

16. Véliz, "Moral Zombies."

17. Cunningham, "'Do Not Hallucinate'"; Wiggers, "Anthropic Publishes the 'System Prompts' That Make Claude Tick."

18. Eamonn Maguire, "Beyond the Hype: Why Superintelligence Will Emerge from Human-AI Orchestration, Not LLMs Alone," *Medium*, October 25, 2025. Another way to phrase the argument is that what makes human cognition superior to that of current AI is that we have the ability to frame things differently (including causally); for more on mental models and frames, see Kenneth Cukier, Viktor Mayer-Schönberger, and Francis de Véricourt's *Framers* (WH Allen, 2021).

19. McGrayne, *Theory That Would Not Die*, 7.

20. Thank you to Nigel Warburton for reminding me of this example. In the original story, which appeared in Bertrand Russell's *Problems of Philosophy* (1912), the turkey is a chicken.

21. Weiser and Schweber, "ChatGPT Lawyer Explains Himself."

22. Sara Merken, "Large US Law Firm Apologizes for AI Errors in Bankruptcy Court Filing," Reuters, October 24, 2025. Evan Gorelick, "Vigilante Lawyers Expose the Rising Tide of A.I. Slop in Court Filings," *The New York Times*, Nov. 7, 2025. For a growing online database of such cases, see: www.damiencharlotin.com/hallucinations.

23. Lawless, "UK Judge Warns of Risk to Justice After Lawyers Cited Fake AI-Generated Cases in Court."

24. Sextus Empiricus, *Outlines of Pyrrhonism*, bk. 2, chap. 15.

25. Hume, *Enquiry Concerning Human Understanding*, 5.1.6/44.

26. For a deeper dive into this rabbit hole, try McGrayne's *Theory That Would Not Die* and Clayton's *Bernoulli's Fallacy*.

27. McGrayne, *Theory That Would Not Die*, 11.

28. Chivers, *Everything Is Predictable*, 53.

29. "A 30% Chance of AI Catastrophe: Samotsvety's Forecasts on AI Risks and the Impact of a Strong AI Treaty," taisc.org/report.

30. Spiegelhalter, *Art of Uncertainty*, 424.

31. "What Are the Chances of an AI Apocalypse?," *Economist*.

32. Crease, "Charles Sanders Peirce and the First Absolute Measurement Standard."

33. Hacking, *Taming of Chance*, 206.

34. Nietzsche, *Daybreak*, 81.

35. Hacking, *Taming of Chance*, 148.

36. Humphreys, "How Erwin Schrödinger Indulged His 'Lolita Complex' in Ireland."

37. Taleb, *Black Swan*, 287.
38. Yates, *How to Expect the Unexpected*, 383–96.

CHAPTER FIVE: WHEN PREDICTIONS BECOME VERDICTS

1. Barton, *Ancient Astrology*, 45; Gibbon, *History of the Decline and Fall of the Roman Empire*, vol. 1.
2. Bezuijen et al., "Pygmalion and Employee Learning"; Eden, "Self-Fulfilling Prophecy as a Management Tool."
3. Arendt, *Origins of Totalitarianism*, 96.
4. Ibid., 457.
5. Rosenthal and Fode, "Effect of Experimenter Bias on the Performance of the Albino Rat."
6. Isen, "How Do You Even Sell a Book Anymore?"
7. The interview is reproduced in Lizzie Gottlieb's documentary, *Turn Every Page* (2022).
8. Deahl, "Is Publishing Too Top-Heavy?"
9. These calculations are made with data retrieved from the records of the antitrust lawsuit between the Department of Justice and Penguin Random House. Madeline McIntosh, then CEO of Penguin Random House, said that she directly approved deals of $1 million or more. Of the $1 billion market of anticipated best-selling books, Penguin Random House accounts for $370 million. McIntosh said she approved "on average" about two hundred such deals, which suggests that a substantial proportion of those books are bought for $1 million or more (assuming that $370 million are used to buy around a third of the twelve hundred books, that would come to about four hundred books in total, half of which are bought for $1 million or more). Admittedly, it might be that the proportion differs in other publishing houses, but the share of PRH is big enough, and the competition with Simon & Schuster on these deals fierce enough, that the numbers are a reasonable assumption. See *Trial*.
10. Irving, "Oxford University's Other Diversity Crisis."
11. Tsang, "Toward a Scientific Inquiry into Superstitious Business Decision-Making."
12. Olson, *Supremacy*, 35.
13. Roberts, *Churchill*.
14. Merton, "Self-Fulfilling Prophecy."
15. Thucydides, *History of the Peloponnesian War*.
16. Rosecrance and Miller, *Next Great War?*
17. Solal and Snellman, "For Female Founders, Fundraising Only from Female VCs Comes at a Cost."
18. Paredes, "Latino Founders Have a Hard Time Raising Money from VCs."
19. Davis, "Funding to Black Founders Was Down in 2023 for the Third Year in a Row."
20. Silver, *On the Edge*, 288.
21. Johnson, *Enough*, 78–79.
22. Morris, *Strength in Numbers*.

23. Earle, "Problems with Polls."

24. Ibid.

25. Rutenberg, Bensinger, and Eder, " 'Red Wave' Washout."

26. Sargent and Tomasky, " 'Red Wave' Redux." Thanks to Roger McNamee for sending me this article.

27. For an example of the ethical dilemmas faced by medical professionals in the context of scarce resources, see Vinay, Baumann, and Biller-Andorno, "Ethics of ICU Triage During Covid-19."

28. For more on the ethics of organ donation, see the work of David Rodríguez-Arias.

29. King and Mertens, "Self-Fulfilling Prophecy in Practical and Automated Prediction."

30. Martinez and Kirchner, "Secret Bias Hidden in Mortgage-Approval Algorithms."

31. Noble, *Algorithms of Oppression*; Buolamwini, *Unmasking AI*; Benjamin, *Race After Technology*.

32. "Molly Russell Inquest: Online Life Was 'the Bleakest of Worlds,' " BBC, Sept. 21, 2022. Also see Marc Silver's documentary film *Molly vs the Machines*.

33. Hacking, *Taming of Chance*, 78.

34. Moynihan, "He Taught Ancient Texts at Oxford."

35. "Oxford Academic Jailed for Child Abuse Images," *BBC News;* Thurston, "Former Oxford Islamic Scholar Convicted of Rape in Switzerland"; Katherine Griffiths, "Oxford's Business School Names Interim Leader After Dean Resigns," *Bloomberg*, Sept. 11, 2025; Marc Ethier, "Oxford Dean Resigns Amid Accusations of Harassment by a Female Colleague," *Poets and Quants*, Sept. 11, 2025. Griffiths, "Oxford University Has Failed Women over Harassment Concerns, Staff Say," Bloomberg. O'Neill, "Oxford Don Quit After Harassment Inquiry," *The Times*. "Degrees of Danger," *Private Eye*, Nov. 27, 2025. Many other cases could be cited on both sides of the Atlantic.

36. Eubanks, *Automating Inequality*.

37. Peters and Matz, "Large Language Models Can Infer Psychological Dispositions of Social Media Users"; Knight, "AI Chatbots Can Guess Your Personal Information." For more on the ways in which tech is based on patterns of exploitation, read Tim Wu's book *The Age of Extraction* (2025).

38. Geddes, "Death of the Legal Subject."

39. Hare, *Technology Is Not Neutral*.

40. Runciman, *Handover*.

41. Véliz, *Ethics of Privacy and Surveillance*.

42. Véliz, *Privacy Is Power;* Sandra Matz, *Mindmasters*. For more on the ways in which tech is based on patterns of exploitation, read Tim Wu's book *The Age of Extraction* (2025).

43. Hetzner, "Larry Ellison Predicts Rise of the Modern Surveillance State Where 'Citizens Will Be on Their Best Behavior.' "

44. R. W. Davies and Stephen G. Wheatcroft, *The Years of Hunger: Soviet Agriculture, 1931–1933* (New York: Palgrave Macmillan, 2004).

45. Lars Daniel, "China's Surveillance State Is Losing Its Grip," *Forbes*, Dec. 5, 2024.

46. Cain, *Perfect Police State*, 3.

47. Ibid., 18.

48. Ibid., 1.

49. Strittmatter, *We Have Been Harmonised,* 100.

50. Ibid., 191.

51. Ibid., 202.

52. Bradley A. Thayer, "For Chinese Firms, Theft of Your Data Is Now a Legal Requirement," *Hill,* Jan. 7, 2021.

53. Cain, *Perfect Police State.*

54. Dann and Haddow, "Just Doing Business or Doing Just Business."

55. Newman, "Apple Is a Chinese Company."

56. Clark and Mozur, "China's Surveillance State Sucks Up Data."

57. Wee, "China Uses DNA to Track Its People, with the Help of American Expertise."

58. Strittmatter, *We Have Been Harmonised,* 188.

59. Kynge et al., "Exporting Chinese Surveillance."

60. Scott, *Seeing Like a State,* 2–5.

61. Major, "Taboo."

62. Dalrymple, "East India Company." For a good book about how corporations interact with authoritarian regimes today, see Anne Applebaum's *Autocracy, Inc.* (Doubleday, 2024).

63. Arendt, *Origins of Totalitarianism,* 444, 457.

64. Ibid., 458.

CHAPTER SIX: THE CRYSTAL BALL IS CRACKED

1. Collins, "Was Jeanne Calment the Oldest Person Who Ever Lived—or a Fraud?" This might have been a false memory of a lie, because Calment's father appears to have been a shipbuilder. Her husband's family, however, did own a dry goods store, and since they were cousins, it's not outlandish that she might have waited on Van Gogh. Arles is not a very big town.

2. Whitney, "In France, a Citizen Turns 120."

3. "120-Year Lease on Life Outlasts Apartment Heir," *New York Times.*

4. Collins, "Was Jeanne Calment the Oldest Person Who Ever Lived—or a Fraud?"

5. Baker, "1,500 Scientists Lift the Lid on Reproducibility."

6. Owens, "Replication Failures in Psychology Not Due to Difference in Study Populations."

7. Glasziou and Chalmers, "Is 85% of Health Research Really 'Wasted'?"

8. Ioannides, "Why Most Published Research Findings Are False."

9. Simmons, Nelson, and Simonsohn, "False-Positive Psychology."

10. Lewis-Kraus, "They Studied Honesty."

11. Keirs, "Why Don't All Researchers Share Their Data Openly?"

12. Watson, "Many Researchers Say They'll Share Data—but Don't."

13. Protzko et al., "Retraction Note: High Replicability of Newly Discovered Social-Behavioural Findings Is Achievable."

14. Lewis-Kraus, "They Studied Honesty." Thanks to Uri Simonsohn for talking with me about these issues.

15. The following works include suggestions of how to fix science: Simmons, Nelson, and Simonsohn, "False-Positive Psychology"; Ioannides, "Why Most Published Research Findings Are False"; Clayton, *Bernoulli's Fallacy;* Romero, "Philosophy of Science and the Replicability Crisis."

16. Axon, "96% of US Users Opt Out of App Tracking in iOS 14.5, Analytics Find."

17. Taleb, *Black Swan,* 109.

18. Messerli, "Chocolate Consumption, Cognitive Function, and Nobel Laureates."

19. Gigerenzer, *How to Stay Smart in a Smart World,* 118. For a great book about how to be a skeptic in a data-driven world, see Carl T. Bergstrom and Jevin D. West's *Calling Bullshit* (Penguin, 2021).

20. Carnap, "On the Application of Inductive Logic."

21. Véliz, *Ethics of Privacy and Surveillance,* chap. 9.

22. Silver, *The Signal and the Noise.*

23. Popper, *Poverty of Historicism,* 124.

24. Gigerenzer, *How to Stay Smart in a Smart World,* 112–17.

25. du Sautoy, *Thinking Better.*

26. Gigerenzer and Brighton, "Homo Heuristicus."

27. Ibid.

28. Aikman et al., "Taking Uncertainty Seriously."

29. Dosi et al., "Rational Heuristics?"

30. Luan, Reb, and Gigerenzer, "Ecological Rationality."

31. "How Better Data Could Lead to Better Sex," *Economist.*

32. Garcia, *Joy of Consent.*

33. Kislin, "Silent Educator"; Harrison and Ollis, "Young People, Pleasure, and the Normalization of Pornography"; Evans, "Porn Effects."

34. Anderson, "End of Theory."

35. Popper, *Poverty of Historicism,* 112.

36. Birhane et al., "Values Encoded in Machine Learning Research." For a book about how big tech is pushing for the value of optimization over and above more important democratic values, see Rob Reich, Mehran Sahami, and Jeremy M. Weinstein's book *System Error* (Harper, 2022). In their book *As If Human* (Yale, 2024), Nigel Shadbolt and Roger Hampson say that we should assess AI decisions as if they came from a human being, which would partly visibilize the values driving those decisions.

37. Edgeworth, "Philosophy of Chance."

38. Silver, *Signal and the Noise,* 45.

39. I take this example, along with the breakdown of the height of two people and the sales of two books, from Nassim Taleb, but I added a zero to the higher salary, both for effect and because it probably reflects today's reality better, almost two decades after the publication of Taleb's book. Taleb, *Black Swan,* 235.

40. Ibid.

41. Thompson, *Merchants of Culture,* 161.

42. According to the transcripts of the trial between Penguin Random House and the Department of Justice, PRH earned 58 percent of its revenue from backlist books in 2020.

43. Grayling, *History of Philosophy*, 297–300.

44. For a great portrait of the Vienna Circle and the Viennese ambience at that time, see Edmonds, *Murder of Professor Schlick*.

45. Popper, *Poverty of Historicism*, xi.

46. Azhar, *Exponential*.

47. Popper, *Poverty of Historicism*, xiii.

48. Ibid., 96.

49. Ibid., 102.

50. Arendt, *Origins of Totalitarianism*, 452.

51. Popper, *Poverty of Historicism*, 147.

52. I call it family legend because when I tried to find the records of a crashed plane around the time my grandmother would've taken it, I found none. My grandmother and great-uncle remembered the episode vividly, so I'm hesitant to dismiss it.

53. Lebow, *Archduke Franz Ferdinand Lives!*

54. I take the last two examples from MacIntyre, *After Virtue*, 116.

55. Plutarch, *Marias* 42.4–5; Barton, *Ancient Astrology*.

56. Tetlock, *Expert Political Judgment*.

57. Beck, *Risk Society*.

58. Milmo, "'Godfather of AI' Shortens Odds of the Technology Wiping Out Humanity over Next 30 Years."

59. Searcey and Kircher, "Torrent of Hate for Health Insurance Industry Follows CEO's Killing."

60. Goldberg, "'Chilling' Fatal Shooting of a CEO Has Business Leaders on Edge."

61. Stanton, "Year Before CEO Shooting, Lawsuit Alleged UHC Used AI to Deny Coverage."

62. Kayser, "Hospitals Are Reporting More Insurance Denials."

63. Cevolini and Esposito, "From Pool to Profile."

64. Herzog, "Algorithmic Bias and Access to Opportunities."

65. Pugh, "Risky Business."

66. Smith and Arnold, "AI Risks Making Some People 'Uninsurable,' Warns UK Financial Watchdog."

67. Clark, "Big Tech Companies Including Intel, Nvidia, and Cisco Were All Infected During the SolarWinds Hack."

68. Kim, "Amazon Server Outage Makes Some Websites Go Dark."

69. Chayka, "Tyranny of the Algorithm"; Chayka, *Filterworld*.

70. Hill, "I Took a 'Decision Holiday' and Put AI in Charge of My Life."

71. Silver, *Signal and the Noise*, 46.

72. Taleb, *Antifragile*.

73. For a criticism of growth, see Raworth, *Doughnut Economics*. For a defense of growth, see Susskind, *Growth*.

74. Rushkoff, "Super-Rich 'Preppers' Planning to Save Themselves from the Apocalypse."
75. Piña, "Mark Zuckerberg Clarifies Reports About His Bunker in Hawaii."
76. Hao, *Empire of AI*, 265.
77. Popper, *Poverty of Historicism*, 146.
78. Seinfeld, *SeinLanguage*, 116.
79. Arendt, *Origins of Totalitarianism*, 188.
80. Ibid., 204.
81. Caro, *Power Broker*, 19.
82. Taleb, *Antifragile*, 307.
83. Rosa, *Uncontrollability of the World*, 111–15.

CHAPTER SEVEN: TRUTH, VIRTUE, AND BEAUTY

1. Wenar, "Deaths of Effective Altruism."
2. Kulish, "How a Scottish Moral Philosopher Got Elon Musk's Number."
3. Visram, "Truth About Elon Musk, Sam Bankman-Fried, and Effective Altruism."
4. McMillan, "How a Fervent Belief Split Silicon Valley—and Fueled the Blowup at OpenAI."
5. For an academic book criticizing effective altruism, see Adams, Crary, and Gruen, *The Good It Promises, the Harm It Does.*
6. The most famous utilitarians (Bentham, Mill, Sidgwick, Hare) have argued that utilitarianism requires us largely to live in accordance with commonsense morality. They make a distinction between the criterion of correctness (what makes an action right) and how we should live. According to this view, utilitarianism tells us which acts are right and wrong, but not how we ought to live. But this move seems like a sleight of hand to me, as it did to other critics like Bernard Williams. I don't think that it is possible for human beings to have two minds in one, and surely if utilitarianism tells us which actions are right and you believe that's true, that will push you to live in accordance with the actions that are right. In my experience, as will become evident, utilitarians do tend to develop a special kind of ruthlessness that makes them the appropriate target of caution. That's not even considering challenges like weakness of will and self-deception.
7. Edmonds, *Parfit*, 152.
8. Foot, "Abortion and the Doctrine of Double Effect"; Thomson, "Killing, Letting Die, and the Trolley Problem."
9. Edmonds, *Parfit*, 83; Parfit, *On What Matters*, xxiii.
10. Sidgwick, *Methods of Ethics*, 489–90. For more on Sidgwick, see Crisp, *Cosmos of Duty.* For a modern defense of secrecy in utilitarianism, see Lazari-Radek and Singer, "Secrecy in Consequentialism."
11. Watson, "Moral Agency"; Christman, "Autonomy in Moral and Political Philosophy."

12. Gettleman, "Meant to Keep Malaria Out, Mosquito Nets Are Used to Haul Fish In."

13. Lewis-Kraus, "They Studied Honesty."

14. In 2023, Nick Bostrom was forced to publish an apology for racist comments he had made in 1996. He had used the n-word and had argued that white people were more intelligent than Black people. Shamefully, in his apology, he didn't retract this argument. He said he didn't have enough information to make that call and that it wasn't his area of expertise. In the wake of the scandal, the Faculty of Philosophy at Oxford declared that it "utterly condemns racism in all its forms." The institute was dissolved in 2024. Is it a coincidence that Nick Bostrom's racist and eugenicist views can be considered direct descendants of predecessors like Galton? I suspect not. Anthony, " 'Eugenics on Steroids.' "

15. A post by Open Philanthropy disclosing grants: "Open Philanthropy recommended two grants totaling £17,324,697 (approximately $23,212,677 at the time of conversion) to the Effective Ventures Foundation (EVF) to support the purchase, renovation, and operation of Wytham Abbey—a venue in Oxford, U.K., for hosting conferences and other events, modeled on similar venues funded by the Rockefeller Foundation and the Brocher Foundation." www.openphilanthropy.org/grants/effective-ventures-foundation-uk-event-venue/.

16. Sarah Emerson, "Effective Altruism–Linked Group Used Millions Donated by FTX to Buy a Czech Castle," *Forbes,* Feb. 28, 2023.

17. Torres, "Fraud, Lies, Exploitation, and Eugenic Fantasies."

18. Ibid.

19. Srinivasan, "Stop the Robot Apocalypse."

20. Singer, "Hinge of History."

21. Nick Bostrom has advanced this argument.

22. Lewis-Kraus, "Reluctant Prophet of Effective Altruism."

23. Véliz, "If You Want to Do the Most Good, Maybe You Shouldn't Work for Wall Street."

24. Alter, "Effective Altruism Promises to Do Good Better."

25. Lewis-Kraus, "Reluctant Prophet of Effective Altruism."

26. Ehrlich and Peterson-Withorn, "Meet the World's Richest 29-Year-Old."

27. Diamond, "Crypto Nerd Sam Bankman-Fried, Who Just Lost $16 Billion, 'Would Never Read a Book' "; Williams, "People Who Don't Read Books."

28. Williams, *Morality.*

29. Silver, *On the Edge,* 373–74.

30. Wenar, "Deaths of Effective Altruism."

31. Williams, *Making Sense of Humanity and Other Philosophical Essays.*

32. Lee-Stronach, "Just Probabilities"; Anderson, "Knowledge and Merely Predictive Evidence."

33. Milmo, "Meta Is Ushering In a 'World Without Facts,' Says Nobel Peace Prize Winner"; Ressa, *How to Stand Up to a Dictator.*

34. Cain, *Perfect Police State,* 142.

35. Strittmatter, *We Have Been Harmonised,* 16–17.

36. Ibid., 18–19.

37. Ibid., 22–23, 39.
38. Ibid., 84.
39. Ibid., 138.
40. Relman, "25 Women Who Have Accused Trump of Sexual Misconduct."
41. Scott, "Supervillain Is the Hero Now."
42. Tolstoy, *Gospel in Brief*, 14.
43. Frans de Waal, "The Surprising Science of Alpha Males," TEDMED, Nov. 2017.
44. Mahdawi, "30 Under 30-Year Sentences."
45. Lewis, *Going Infinite*, chap. 8.
46. Ibid.
47. Silver, *On the Edge*, 391.
48. McBrayer, "Why Our Children Don't Think There Are Moral Facts."

CHAPTER EIGHT: DEFYING THE ODDS

1. Donald, "I Reported Harassment and Was Silenced—and I'm a Senior Academic"; Williams and Massinger, "How Women Are Harassed Out of Science."
2. Lytton, "AI Hiring Tools May Be Filtering Out the Best Job Applicants."
3. Carr, "Giving Viewers What They Want."
4. Plummer, "This Is How Netflix's Top-Secret Recommendation System Works."
5. Sackett and Henry, *Seinfeld: How It Began*.
6. Battaglio, "Biz: The Research Memo That Almost Killed Seinfeld."
7. Tett, "Can AI Crack Comedy?"
8. Of course, there is also cruel, sexist, and racist humor, among other kinds of toxic humor. Some of the most chilling "humor" I've heard of, if you can call it that, is the ridiculing of victims of crimes or wars. For more on the morality of humor, see Shoemaker, *Wisecracks*.
9. Some of the next few paragraphs first appeared in Véliz, "If AI Is Predicting Your Future, Are You Still Free?"
10. Hacking, *Taming of Chance*, 114.
11. Durkheim, *Suicide*, 273.
12. Hawking, *Black Holes and Baby Universes and Other Essays*, 122.
13. Barton, *Ancient Astrology*, 38.
14. Magnuson, *For Profit*, 9.
15. Sommerlad, "Silicon Six Accused of Avoiding Nearly $300Bn in Taxes, New Report Says."
16. Schaake, "Lobbying for Unfettered Innovation Is Bad for Democracy."
17. Magnuson, *For Profit*, 38, 304–5.
18. Surowiecki, *Wisdom of Crowds*.
19. Popper, *Poverty of Historicism*, 147.
20. Hacking, *Taming of Chance*, 36.
21. Ibid., 37.
22. Ibid., 130.
23. Cutler and Reber, "Paying for Health Insurance."
24. Einav, Finkelstein, and Fisman, *Risky Business*.

25. L. Slater et al., "Global Changes in 20-Year, 50-Year, and 100-Year River Floods," *Geophysical Research Letters* 48, no. 6 (2021): e2020GL091824.
26. The Delta Act, wetten.overheid.nl/BWBR0002283/2004-07-01; The Dutch Meteorological Institute, www.knmi.nl/kennis-en-datacentrum/achtergrond/stormklimaat-en-hoogwaters.
27. Heffernan, *Uncharted*, 84.
28. Harford, "Forecast Mostly Gloomy, but That's a Good Thing."
29. For more on scenario planning, see the work of my colleague at Oxford Rafael Ramírez and that of Amy Webb and Diana Mangalagiu, among many others.

CHAPTER NINE: HOW TO THRIVE AMID UNCERTAINTY

1. Thanks to Tara Westover for reminding me of this quotation.
2. William James, *The Principles of Psychology*.
3. An apocryphal quotation, unfortunately.
4. Flaminia Luck, "Nightclub Bouncer Behind £12M Plot to Reveal Formula One Star Michael Schumacher's Health Secrets Jailed," LBC, Feb. 12, 2025.
5. The technical term for unintentional harm caused by doctors is "iatrogenesis." As Taleb puts it, "If you want to accelerate someone's death, give him a personal doctor." Taleb, *Antifragile*.
6. van der Horst, "Ancient Jewish Bibliomancy."
7. Some writers go as far as arguing that it's all fiction (it's not!). For a history of fiction that takes such an interpretation, see Volpi, *La invención de todas las cosas*.
8. Alex Blasdel, "Days of the Jackal: How Andrew Wylie Turned Serious Literature into Big Business," *Guardian*, Nov. 9, 2023.
9. Stone, "Amazon Erases Orwell Books from Kindle"; Ugwu, "It's Their Content, You're Just Licensing It."
10. Nguyen, "Peek Inside Robert Caro's Home Library, Hidden Shelves and All."
11. Burgess, "All the Ways Amazon Tracks You—and How to Stop It"; Gartenberg, "Why Amazon Is Tracking Every Time You Tap Your Kindle."
12. Richmond, "Printed Book Is Doomed."
13. Plato, *Theaetetus* 174a.
14. Aristotle, *Politics*.
15. Annas, *Morality of Happiness*, 340–41.
16. Fields, "False Prophets and Fake Prophecies in Lucian."
17. Ehrenfeld, "Why Epicurean Ideas Suit the Challenges of Modern Secular Life."
18. Rodríguez, "El hombre más rico del mundo no duerme mejor que tú."
19. Hearing, "Elon Musk's Warning to Leaders Wanting to Be Like Him."
20. Seneca, *On the Shortness of Life*.
21. Arendt, *Eichmann in Jerusalem*, 276.
22. "Global Leaders Educated at Oxford," Blavatnik School of Government, University of Oxford, June 23, 2011, www.bsg.ox.ac.uk.

EPILOGUE: TEN LESSONS IN PREDICTION

1. Seneca, *On the Shortness of Life*.

POSTSCRIPT : AI ETHICS 101: DIGITAL TECH'S ORIGINAL SINS

1. Véliz, "Three Things Digital Ethics Can Learn from Medical Ethics."

APPENDIX : ANCIENT FORMS OF DIVINATION

1. Aristotle, *On Divination Through Dreams.*
2. Rochberg, *Heavenly Writing,* 68.
3. Johnston, *Ancient Greek Divination,* 161.
4. Although I'm focusing on Western divination practices, it's worth mentioning that the idea of people channeling spirits for the purposes of prediction is found in many other cultures. In India, *theyyam* is a ritual whereby priests can invite a deity to possess them so they can answer devotees' questions. In Tibetan culture, *kuten* get likewise possessed by spirits who can issue prophecies. Tibetan descriptions of oracular possession are, at least in some ways, curiously similar to ancient Greek ones. Tibetan oracular ceremonies begin with chanted invocations and prayers, accompanied by horns, cymbals, and drums. When the *kuten* enters his trance, he becomes capable of incredible physical feats. A skinny man who before the trance could barely walk wearing an elaborate outfit that weighs more than seventy pounds is suddenly able to dance in it after being crowned by a helmet weighing thirty pounds more, all the while wielding a heavy sword. His breath becomes agitated; his face reddens and becomes covered with sweat. He huffs and puffs while he answers questions. Finally, he collapses in complete exhaustion, signifying the end of the possession, his attendants carrying him away, lifeless. There are a few documentaries and online footage documenting this otherworldly ritual. Gyatso, *Freedom in Exile,* 211–14.
5. Barton, *Ancient Astrology,* 186.
6. Pliny, *Natural History* 28, 29.

SELECTED BIBLIOGRAPHY

Adams, Carol J., Alice Crary, and Lori Gruen, eds. *The Good It Promises, the Harm It Does: Critical Essays on Effective Altruism.* Oxford: Oxford University Press, 2023.

Aikman, David, Mirta Galesic, Gerd Gigerenzer, Sujit Kapadia, Konstantinos Katsikopoulos, Amit Kothiyal, Emma Murphy, and Tobias Neumann. "Taking Uncertainty Seriously: Simplicity Versus Complexity in Financial Regulation." *Industrial and Corporate Change* 30, no. 2 (2021): 317–45.

Alter, Charlotte. "Effective Altruism Promises to Do Good Better. These Women Say It Has a Toxic Culture of Sexual Harassment and Abuse." *Time,* Feb. 3, 2023.

Ambady, Nalini. "The Perils of Pondering: Intuition and Thin Slice Judgments." *Psychological Inquiry* 21 (2010): 271–78.

Anderson, Chris. "The End of Theory: The Data Deluge Makes the Scientific Method Obsolete." *Wired,* Junc 23, 2008.

Anderson, Haley Schilling. "Knowledge and Merely Predictive Evidence." *Philosophical Studies* 182 (2025): 467–85.

Angwin, Julia, Jeff Larson, Charlie Savage, James Risen, Henrik Moltke, and Laura Poitras. "NSA Spying Relies on AT&T's 'Extreme Willingness to Help.'" ProPublica, Aug. 15, 2015.

Annas, Julia. *The Morality of Happiness.* Oxford: Oxford University Press, 1993.

Anscombe, G. E. M. "Aristotle and the Sea Battle." *Mind* 65, no. 257 (1956): 1–15.

Anthony, Andrew. "'Eugenics on Steroids': The Toxic and Contested Legacy of Oxford's Future of Humanity Institute." *Guardian,* April 28, 2024.

Applebaum, Anne. *Autocracy, Inc.* New York: Doubleday, 2024.

Arendt, Hannah. *Eichmann in Jerusalem: A Report on the Banality of Evil.* London: Penguin, 2006.

———. *The Origins of Totalitarianism.* London: Penguin, 2017.

Axon, Samuel. "96% of US Users Opt Out of App Tracking in iOS 14.5, Analytics Find." *Ars Technica,* May 7, 2021.

Ayers, J. W., A. Poliak, M. Dredze, E. C. Leas, Z. Zhu, J. B. Kelley, D. J. Faix, et al. "Comparing Physician and Artificial Intelligence Chatbot Responses to Patient Questions Posted to a Public Social Media Forum." *JAMA Internal Medicine* 183, no. 6 (2023): 589–96.

Azhar, Azeem. *Exponential: How Accelerating Technology Is Leaving Us Behind and What to Do About It.* London: Random House, 2021.

Baker, Monya. "1,500 Scientists Lift the Lid on Reproducibility." *Nature* 533 (2016): 452–54.

Barnett, Kendra. "Record-Breaking 2024 US Political Ad Spend Likely to Disrupt Commercial Media Landscape." *Drum,* Aug. 21, 2024.

Barton, Tamsyn. *Ancient Astrology.* London: Routledge, 1994.

Battaglio, Stephen. "The Biz: The Research Memo That Almost Killed Seinfeld." *TV Guide,* June 27, 2014.

BBC News. "Facebook Stops Sending Staff to Help Political Campaigns." Sept. 21, 2018.

———. "Oxford Academic Jailed for Child Abuse Images." March 19, 2020.

Beck, Ulrich. *Risk Society: Towards a New Modernity.* Translated by Mark Ritter. London: Sage, 1992.

Bender, Emily M., and Alex Hanna. *The AI Con: How to Fight Big Tech's Hype and Create the Future We Want.* London: Bodley Head, 2025.

Benjamin, Ruha. *Race After Technology.* Cambridge, U.K.: Polity, 2019.

Bergstrom, Carl T., and Jevin D. West. *Calling Bullshit.* New York: Penguin, 2021.

Bernstein, Peter L. *Against the Gods: The Remarkable Story of Risk.* New York: Wiley, 1998.

Bezuijen, Xander M., Peter T. van den Berg, Karen van Dam, and Henk Thierry. "Pygmalion and Employee Learning: The Role of Leader Behaviors." *Journal of Management* 35, no. 5 (2009).

Bickel, Eric J., and Seong Dae Kim. "Verification of the Weather Channel Probability of Precipitation Forecasts." *Monthly Weather Review* 136, no. 12 (2008): 4867–81.

Biddle, Sam. "How Peter Thiel's Palantir Helped the NSA Spy on the Whole World." *Intercept,* Feb. 22, 2017.

Birhane, Abeba. "The Impossibility of Automating Ambiguity." *Artificial Life* 27 (2021): 44–61.

Birhane, Abeba, Pratyusha Kalluri, Dallas Card, William Agnew, Ravit Dotan, and Michelle Bao. "The Values Encoded in Machine Learning Research." *FAccT '22: Proceedings of the 2022 ACM Conference on Fairness, Accountability, and Transparency* (2022): 173–84.

Buolamwini, Joy. *Unmasking AI.* London: Random House, 2024.

Burgess, Matt. "All the Ways Amazon Tracks You—and How to Stop It." *Wired,* June 22, 2021.

Burns, Dasha, and Holly Otterbein. "Trump Staff 'Furious' After Musk Trashes AI Project." *Politico,* Jan. 23, 2025.

Cain, Geoffrey. *The Perfect Police State: An Undercover Odyssey into China's Terrifying Surveillance Dystopia of the Future.* New York: PublicAffairs, 2021.

Carnap, Rudolf. "On the Application of Inductive Logic." *Philosophy and Phenomenological Research* 8 (1947): 133–48.

Caro, Robert. *The Power Broker: Robert Moses and the Fall of New York.* New York: Vintage, 1975.

Carr, David. "Giving Viewers What They Want." *New York Times,* Feb. 24, 2013.

Casselman, Ben. "Economists Are in the Wilderness. Can They Find a Way Back to Influence?" *New York Times,* Jan. 10, 2025.

Cevolini, Alberto, and Elena Esposito. "From Pool to Profile—Social Consequences of Algorithmic Prediction in Insurance." *Big Data & Society* 1 (2020).

Chayka, Kyle. *Filterworld: How Algorithms Flattened Culture.* New York: Doubleday, 2024.

———. "The Tyranny of the Algorithm: Why Every Coffee Shop Looks the Same." *Guardian,* Jan. 16, 2024.

Chelli, M., J. Descamps, V. Lavoue, C. Trojani, M. Azar, M. Deckert, J. L. Raynier, et al. "Hallucination Rates and Reference Accuracy of ChatGPT and Bard for Systematic Reviews: Comparative Analysis." *Journal of Medical Internet Research* 26 (2024): e53164.

Chevalier, Michel. "Opening Address." *Journal de la Société de Statistique de Paris* 1 (1860): 1–6.

Chivers, Tom. *Everything Is Predictable.* London: Weidenfeld & Nicolson, 2024.

Christman, John. "Autonomy in Moral and Political Philosophy." In *The Stanford Encyclopedia of Philosophy,* edited by Edward N. Zalta. Stanford University, 2015.

Clark, Don, and Paul Mozur. "China's Surveillance State Sucks Up Data. US Tech Is Key to Sorting It." *New York Times,* Nov. 22, 2020.

Clark, Mitchell. "Big Tech Companies Including Intel, Nvidia, and Cisco Were All Infected During the SolarWinds Hack." *Verge,* Dec. 21, 2020.

Clayton, Aubrey. *Bernoulli's Fallacy: Statistical Illogic and the Crisis of Modern Science.* New York: Columbia University Press, 2021.

Collins, Lauren. "Was Jeanne Calment the Oldest Person Who Ever Lived—or a Fraud?" *New Yorker,* Feb. 10, 2020.

Crawford, Kate. *Atlas of AI.* New Haven, Conn.: Yale University Press, 2021.

Crease, Robert P. "Charles Sanders Peirce and the First Absolute Measurement Standard." *Physics Today* 62, no. 12 (2009): 39–44.

Crisp, Roger. *The Cosmos of Duty: Henry Sidgwich's Methods of Ethics.* Oxford: Oxford University Press, 2015.

Cukier, Kenneth, Viktor Mayer-Schönberger, and Francis de Véricourt. *Framers.* London: WH Allen, 2021.

Cunningham, Andrew. "'Do Not Hallucinate': Testers Find Prompts Meant to Keep Apple Intelligence on the Rails." *Ars Technica,* Aug. 6, 2024.

Cutler, David M., and Sarah J. Reber. "Paying for Health Insurance: The Trade-Off Between Competition and Adverse Selection." *Quarterly Journal of Economics* 113 (1998): 433–66.

Dahms, Thomas. "Diligent Bureaucrats and the Expulsion of Jews from West Prussia, 1772–1786." *German History* 39, no. 3 (2021): 335–57.

Dalrymple, William. "The East India Company: The Original Corporate Raiders." *Guardian,* March 4, 2015.

Daniel, Caroline, and Maija Palmer. "Google's Goal: To Organise Your Daily Life." *Financial Times,* May 22, 2007.

Dann, Elijah G., and Neil Haddow. "Just Doing Business or Doing Just Business: Google, Microsoft, Yahoo!, and the Business of Censoring China's Internet." *Journal of Business Ethics* 79, no. 3 (2008): 219–34.

Davis, Dominic-Madori. "Funding to Black Founders Was Down in 2023 for the Third Year in a Row." *TechCrunch,* Jan. 17, 2024.

Deahl, Rachel. "Is Publishing Too Top-Heavy?" *Publishers Weekly,* Nov. 1, 2019.

"The Decriminalisation of Rape." Centre for Women's Justice, End Violence Against Women Coalition, Imkaan, and Rape Crisis England & Wales, 2020.

Diamond, Jonny. "Crypto Nerd Sam Bankman-Fried, Who Just Lost $16 Billion, 'Would Never Read a Book.'" *LitHub,* Nov. 11, 2022.

Diente, Tanya. "Princess Diana 'Predicted' She Would Die in a Staged Car Accident, Says Advisor." *International Business Times,* Aug. 19, 2022.

Donald, Athene. "I Reported Harassment and Was Silenced—and I'm a Senior Academic." *Guardian,* Oct. 8, 2018.

Dosi, Giovanni, Mauro Napoletano, Andrea Roventini, Joseph E. Stiglitz, and Tania Treibich. "Rational Heuristics? Expectations and Behaviors in Evolving Economies with Heterogeneous Interacting Agents." *Economic Inquiry* 58, no. 3 (2020): 1487–516.

Duffy, Clare, and Amy O'Kruk. "How the Feud Between Elon Musk and Donald Trump Exploded over 72 Hours." CNN, June 6, 2025.

Dunnington, Waldo G. *Carl Friedrich Gauss: Titan of Science.* Mathematical Association of America, 2004.

Durkheim, Émile. *Suicide: A Study in Sociology.* London: Routledge, 2005.

du Sautoy, Marcus. *Thinking Better: The Art of the Shortcut.* London: 4th Estate, 2021.

Earle, Samuel. "The Problems with Polls." *New York Review of Books,* Oct. 17, 2024.

Economist. "Astrologers Are Predicting the Result of America's Election." Sept. 9, 2024.

———. "How Better Data Could Lead to Better Sex." Dec. 19, 2024.

———. "What Are the Chances of an AI Apocalypse?" July 10, 2023.

Eden, Dov. "Self-Fulfilling Prophecy as a Management Tool: Harnessing Pygmalion." *Academy of Management Review* 9, no. 1 (1984): 64–73.

Edgeworth, F. Y. "The Philosophy of Chance." *Mind* 31, no. 123 (1922).

Edmonds, David. *The Murder of Professor Schlick: The Rise and Fall of the Vienna Circle.* Princeton, N.J.: Princeton University Press, 2020.

———. *Parfit: A Philosopher and His Mission to Save Morality.* Princeton, N.J.: Princeton University Press, 2023.

Ehrenfeld, Temma. "Why Epicurean Ideas Suit the Challenges of Modern Secular Life." *Aeon,* July 19, 2019.

Ehrlich, Steven, and Chase Peterson-Withorn. "Meet the World's Richest 29-Year-Old: How Sam Bankman-Fried Made a Record Fortune in the Crypto Frenzy." *Forbes,* Oct. 6, 2021.

Einav, Liran, Amy Finkelstein, and Ray Fisman. *Risky Business.* New Haven, Conn.: Yale University Press, 2022.

Eubanks, Virginia. *Automating Inequality.* New York: Picador, 2018.

Evans, Max. "Porn Effects: 'My Expectations of Sex and Body Image Were Warped.'" *BBC News,* Oct. 24, 2020.

Ferens, Wayne. "The Edsel: An Entirely New Kind of Car." *Motorcities,* Jan. 3, 2024.

Field, Matthew. "Doctors Fear ChatGPT Is Fuelling Psychosis." *Telegraph,* July 27, 2025.

Fields, Dana. "False Prophets and Fake Prophecies in Lucian." In *Divination and Prophecy in the Ancient Greek World,* edited by Roger D. Woodard. Cambridge, U.K.: Cambridge University Press, 2023.

Foot, Philippa. "Abortion and the Doctrine of Double Effect." *Oxford Review* 5 (1967): 28–41.

Forst, Rainer. "Noumenal Power." *Journal of Political Philosophy* 23, no. 2 (2015): 111–27.

Fourny, Marc. "Élizabeth Teissier raconte ses rendez-vous avec Mitterrand et Chirac." *Le point,* April 5, 2022.

Frankfurt, Harry. *On Bullshit.* Princeton, N.J.: Princeton University Press, 2005.

Frenkel, Sheera. "The Militarization of Silicon Valley." *New York Times,* Aug. 5, 2025.

Frenkel, Sheera, and Aaron Krolik. "Trump Taps Palantir to Compile Data on Americans." *New York Times,* May 30, 2025.

Friedman, Walter A. *Fortune Tellers: The Story of America's First Economic Forecasters.* Princeton, N.J.: Princeton University Press, 2014.

Galileo. *Discoveries and Opinions of Galileo.* New York: Doubleday, 1957.

Galton, Francis. "Regression Towards Mediocrity in Hereditary Stature." *Journal of the Anthropological Institute of Great Britain and Ireland* 15 (1886): 246–63.

Garcia, Manon. *The Joy of Consent.* Cambridge, Mass.: Belknap Press of Harvard University Press, 2023.

Gartenberg, Chaim. "Why Amazon Is Tracking Every Time You Tap Your Kindle." *Verge,* Jan. 31, 2020.

Geddes, Katrina. "The Death of the Legal Subject." *Vanderbilt Journal of Entertainment and Technology Law* 25, no. 1 (2023).

Gellman, Barton. *Dark Mirror.* London: Bodley Head, 2020.

Gerstein, Josh. "The Anti-Democratic Worldview of Steve Bannon and Peter Thiel." *Politico,* March 29, 2020.

Gettier, Edmund. "Is Justified True Belief Knowledge." *Analysis* 23, no. 6 (1963): 121–23.

Gettleman, Jeffrey. "Meant to Keep Malaria Out, Mosquito Nets Are Used to Haul Fish In." *New York Times,* Jan. 24, 2015.

Gibbon, Edward. *The History of the Decline and Fall of the Roman Empire.* Vol. 1. Project Gutenberg, 2021.

Gigerenzer, Gerd. *How to Stay Smart in a Smart World.* London: Allen Lane, 2021.

Gigerenzer, Gerd, and Henry Brighton. "Homo Heuristicus: Why Biased Minds Make Better Inferences." *Topics in Cognitive Science* 1, no. 1 (2009): 107–43.

Gigerenzer, Gerd, and David J. Murray. *Cognition as Intuitive Statistics.* Hillsdale, N.J.: Erlbaum, 1987.

Gillispie, Charles Coulston. *Pierre-Simon Laplace, 1749–1827: A Life in Exact Science.* Princeton, N.J.: Princeton University Press, 1997.

Glasziou, Paul, and Iain Chalmers. "Is 85% of Health Research Really 'Wasted'?" *British Medical Journal Opinion,* Jan. 14, 2016.

Goldberg, Emma. "The 'Chilling' Fatal Shooting of a CEO Has Business Leaders on Edge." *New York Times,* Dec. 6, 2024.

Graeber, David. *Debt: The First 5,000 Years.* Brooklyn: Melville House, 2014.

Grayling, A. C. *The History of Philosophy.* London: Penguin, 2019.

Griffiths, Ben, and Graeme Culliford. "Princess Diana's Astrologer Read Her Horoscope to Her a Month Before She Died." *Sun,* Aug. 27, 2017.

Griffiths, Katherine. "Oxford University Has Failed Women over Harassment Concerns, Staff Say." *Bloomberg,* Nov. 19, 2025.

Guyer, Jonathan. "Big Tech's Push into Military AI Is Troubling." *Financial Times,* June 25, 2025.

Gyatso, Tenzin. *Freedom in Exile: The Autobiography of the Dalai Lama.* New York: Cornelia & Michael Bessie, 1990.

Hacking, Ian. *The Emergence of Probability.* Cambridge, U.K.: Cambridge University Press, 2006.

———. *The Taming of Chance.* Cambridge, U.K.: Cambridge University Press, 2010.

Hao, Karen. *Empire of AI: Dreams and Nightmares in Sam Altman's OpenAI.* New York: Penguin Press, 2025.

Harford, Tim. "An Astonishing Record—of Complete Failure." *Financial Times,* May 30, 2014.

———. "Forecast Mostly Gloomy, but That's a Good Thing." *Financial Times,* Jan. 3, 2025.

Harrison, Lyn, and Debbie Ollis. "Young People, Pleasure, and the Normalization of Pornography: Sexual Health and Well-Being in a Time of Proliferation?" In *Handbook of Children and Youth Studies,* edited by Johanna Wyn and Helen Cahill. Singapore: Springer, 2015.

Hawking, Stephen. *Black Holes and Baby Universes and Other Essays.* London: Bantam, 1994.

Hearing, Alice. "Elon Musk's Warning to Leaders Wanting to Be Like Him: 'Be Careful What You Wish For. The Amount I Torture Myself Is Next-Level.'" *Forbes,* Nov. 14, 2022.

Heaven, Will Douglas. "Hundreds of AI Tools Have Been Built to Catch Covid. None of Them Helped." *MIT Technology Review,* July 30, 2021.

Heffernan, Margaret. *Uncharted: How Uncertainty Can Power Change.* London: Simon & Schuster, 2020.

Heinecke, Liz. "When Marie Curie Was Almost Excluded from Winning the Nobel Prize." *LitHub,* Feb. 18, 2021.

Henshall, Will. "There's an AI Lobbying Frenzy in Washington. Big Tech Is Dominating." *Time,* April 30, 2024.

Herzog, Lisa. "Algorithmic Bias and Access to Opportunities." In *The Oxford Handbook of Digital Ethics,* edited by Carissa Véliz, 413–32. Oxford: Oxford University Press, 2023.

Hetzner, Christiaan. "Larry Ellison Predicts Rise of the Modern Surveillance State Where 'Citizens Will Be on Their Best Behavior.'" *Forbes,* Sept. 17, 2024.

Hill, Kashmir. "I Took a 'Decision Holiday' and Put AI in Charge of My Life." *New York Times,* Nov. 1, 2024.

———. "Why Is ChatGPT Telling People to Email Me?" *New York Times,* June 29, 2025.

———. "Your Car Is Tracking You. Abusive Partners May Be, Too." *New York Times,* Dec. 31, 2023.

Hsee, Christopher K., and Reid Hastie. "Decision and Experience: Why Don't We Choose What Makes Us Happy?" *Trends in Cognitive Science* 10, no. 1 (2006): 31–37.

Hume, David. *An Enquiry Concerning Human Understanding.* Edited by T. L. Beauchamp. Oxford: Clarendon Press, 2000.

Humphreys, Joe. "How Erwin Schrödinger Indulged His 'Lolita Complex' in Ireland." *Irish Times,* Dec. 11, 2021.

Ioannides, John P. A. "Why Most Published Research Findings Are False." *PloS Medicine* 19, no. 8 (2005).

Irving, Emma. "Oxford University's Other Diversity Crisis." *Economist, 1843 Magazine,* March 1, 2023.

Isen, Tajja. "How Do You Even Sell a Book Anymore?" *Walrus,* Dec. 18, 2023.

Jaser, Zahira, and Dimitra Petrakaki. "Are You Prepared to Be Interviewed by an AI?" *Harvard Business Review,* Feb. 7, 2023.

Jenkins, Holman W. "Google and the Search for the Future." *Wall Street Journal,* Aug. 14, 2010.

Johnson, Harriet. *Enough: The Violence Against Women and How to End It.* London: William Collins, 2022.

Johnston, Sarah Iles. *Ancient Greek Divination.* West Sussex: Wiley-Blackwell, 2008.

Kay, Richard. "The Spare Ribs of Dante's Michael Scot." *Dante Studies* 103 (1985): 1–14.

Kayser, Alexis. "Hospitals Are Reporting More Insurance Denials. Is AI Driving Them?" *Newsweek,* Nov. 13, 2024.

Keirs, Daniel. "Why Don't All Researchers Share Their Data Openly?" *Times Higher Education,* Oct. 25, 2024.

Kim, Chloe. "Amazon Server Outage Makes Some Websites Go Dark." BBC, June 13, 2023.

King, Owen C., and Mayli Mertens. "Self-Fulfilling Prophecy in Practical and Automated Prediction." *Ethical Theory and Moral Practice* 26 (2023): 127–52.

Kislin, Nancy J. "Silent Educator: The Impact of Porn on Young Minds." *Psychology Today,* Aug. 21, 2023.

Klee, Miles. "He Had a Mental Breakdown Talking to ChatGPT. Then Police Killed Him." *Rolling Stone,* June 22, 2025.

———. "People Are Losing Loved Ones to AI-Fueled Spiritual Fantasies." *Rolling Stone,* May 4, 2025.

Knight, Will. "AI Chatbots Can Guess Your Personal Information." *Wired,* Oct. 17, 2023.

Kolata, Gina. "When Doctors Use a Chatbot to Improve Their Bedside Manner." *New York Times,* June 12, 2023.

Kraaijeveld, S. R., and T. Sharon. "The Increasing Influence of Big Tech in Health and Medicine and the Need for a Public Health Ethics Perspective." *Public Health Ethics* 18, no. 2 (2025): phaf005.

Krause, Carsten. "Case Study: Amazon's AI-Driven Supply Chain: A Blueprint for the Future of Global Logistics." *CDO Times,* Aug. 23, 2024.

Kulish, Nicholas. "How a Scottish Moral Philosopher Got Elon Musk's Number." *New York Times,* Oct. 8, 2022.

Kynge, James, Valerie Hopkins, Helen Warrell, and Kathrin Hille. "Exporting Chinese Surveillance: The Security Risks of 'Smart Cities.'" *Financial Times,* June 9, 2021.

Laplace, Pierre-Simon. *A Philosophical Essay on Probabilities.* Translated by F. W. Truscott and F. L. Emory. London: Chapman & Hall, 1902.

Lawless, Jill. "UK Judge Warns of Risk to Justice After Lawyers Cited Fake AI-Generated Cases in Court." *Independent,* June 7, 2025.

Lazari-Radek, Katarzyna de, and Peter Singer. "Secrecy in Consequentialism: A Defence of Esoteric Morality." *Ratio* 23 (2019): 34–58.

Lebow, Richard Ned. *Archduke Franz Ferdinand Lives! A World Without World War I.* New York: Palgrave Macmillan, 2014.

Lee-Stronach, Chad. "Just Probabilities." *Noûs* 58 (2024): 948–72.

Lewis, H. A. G. "Reviewed Work: The Crime of Claudius Ptolemy by Robert R. Newton." *Imago Mundi* 31 (1979): 105–7.

Lewis, Michael. *Going Infinite.* London: Penguin, 2023.

Lewis-Kraus, Gideon. "The Reluctant Prophet of Effective Altruism." *New Yorker,* Aug. 8, 2022.

———. "They Studied Honesty. Was Their Work a Lie?" *New Yorker,* Sept. 30, 2023.

Luan, Shenghua, Jochen Reb, and Gerd Gigerenzer. "Ecological Rationality: Fast-and-Frugal Heuristics for Managerial Decision Making Under Uncertainty." *Academy of Management Journal* 62, no. 6 (2019): 1735–59.

Lytton, Charlotte. "AI Hiring Tools May Be Filtering Out the Best Job Applicants." BBC, Feb. 16, 2024.

MacIntyre, Alasdair. *After Virtue.* London: Bloomsbury, 2007.

Macintyre, Ben. "Precrime Profiling Is No Longer a Fantasy." *Sunday Times,* Aug. 1, 2025.

Magnuson, William. *For Profit: A History of Corporations.* London: Basic Books, 2022.

Mahdawi, Arwa. "30 Under 30-Year Sentences: Why So Many of Forbes' Young Heroes Face Jail." *Guardian,* April 7, 2023.

Major, Andrea. "Taboo: The East India Company and the True Horrors of Empire." *Conversation,* Feb. 24, 2017.

Martinez, Emmanuel, and Lauren Kirchner. "The Secret Bias Hidden in Mortgage-Approval Algorithms." *Markup,* Aug. 25, 2021.

Matsumi, Hideyuki, and Daniel J. Solove. "The Prediction Society: AI and the Problems of Forecasting the Future." *University of Illinois Law Review* 1 (2025).

Matz, Sandra. *Mindmasters: The Data-Driven Science of Predicting and Changing Human Behavior.* Cambridge: Harvard Business Review, 2025.

McBrayer, Justin P. "Why Our Children Don't Think There Are Moral Facts." *New York Times,* March 2, 2015.

McGrayne, Sharon Bertsch. *The Theory That Would Not Die: How Bayes' Rule Cracked the Enigma Code, Hunted Down Russian Submarines & Emerged Triumphant from Two Centuries of Controversy.* New Haven, Conn.: Yale University Press, 2011.

McMillan, Robert. "How a Fervent Belief Split Silicon Valley—and Fueled the Blowup at OpenAI." *Wall Street Journal,* Nov. 22, 2023.

Merton, Robert K. "The Self-Fulfilling Prophecy." *Antioch Review* 8, no. 2 (1948): 193–210.

Messerli, Franz H. "Chocolate Consumption, Cognitive Function, and Nobel Laureates." *New England Journal of Medicine* 367, no. 16 (2012): 1562–64.

Milmo, Dan. "'Godfather of AI' Shortens Odds of the Technology Wiping Out Humanity over Next 30 Years." *Guardian,* Dec. 27, 2024.

———. "Meta Is Ushering In a 'World Without Facts,' Says Nobel Peace Prize Winner." *Guardian,* Jan. 8, 2025.

Mollick, Ethan. *Co-Intelligence: Living and Working with AI.* London: W. H. Allen, 2024.

Moore, Elaine. "Blame AI for Your Gruelling Job Interview." *Financial Times,* Dec. 4, 2024.

Morris, G. Elliott. *Strength in Numbers: How Polls Work and Why We Need Them.* London: W. W. Norton, 2023.

Moynihan, Colin. "He Taught Ancient Texts at Oxford. Now He Is Accused of Stealing Some." *New York Times,* Sept. 24, 2021.

Muller, Jerry Z. *The Tyranny of Metrics.* Princeton, N.J.: Princeton University Press, 2018.

Newman, Jay. "Apple Is a Chinese Company." *Financial Times,* May 3, 2023.

New York Times. "A 120-Year Lease on Life Outlasts Apartment Heir." Dec. 29, 1995.

Ng, Alfred. "Google Is Giving Data to Police Based on Search Keywords, Court Docs Show." *CNET,* Oct. 8, 2020.

Nguyen, Sophia. "A Peek Inside Robert Caro's Home Library, Hidden Shelves and All." *Washington Post,* Jan. 17, 2025.

Nietzsche, Friedrich. *Daybreak.* Edited by Maudemarie Clark and Brian Leiter. Cambridge, U.K.: Cambridge University Press, 2012.

Noble, Safiya. *Algorithms of Oppression: How Search Engines Reinforce Racism.* New York: New York University Press, 2018.

"Oath of Hippocrates." In *Encyclopedia of Bioethics,* edited by W. T. Reich, 2632. New York: Macmillan, 1995.

Obermeyer, Z., B. Powers, C. Vogeli, and S. Mullainathan. "Dissecting Racial Bias in an Algorithm Used to Manage the Health of Populations." *Science* 366, no. 6464 (2019): 447–53.

Olomi, Ali A. "Baghdad: The Once and Future City of Stars." *Public Books,* Feb. 8, 2022.

Olson, Parmy. *Supremacy: AI, ChatGPT, and the Race That Will Change the World.* London: Macmillan, 2024.

O'Neill, Sean. "Oxford Don Quit After Harassment Inquiry." *The Times,* Nov. 17, 2025.

Owens, Brian. "Replication Failures in Psychology Not Due to Difference in Study Populations." *Nature,* Nov. 19, 2018.

Ozalp, Hakan, Pinar Ozcan, Dize Dinckol, Markos Zachariadis, and Annabelle Gawer. "'Digital Colonization' of Highly Regulated Industries: An Analysis of Big Tech Platforms' Entry into Health Care and Education." *California Management Review* 64, no. 4 (2022).

Paredes, Arielle. "Latino Founders Have a Hard Time Raising Money from VCs." *Wired,* Jan. 26, 2022.

Parfit, Derek. *On What Matters.* Oxford: Oxford University Press, 2011.

Peters, Heinrich, and Sandra Matz. "Large Language Models Can Infer Psychological Dispositions of Social Media Users." *PNAS Nexus* 3, no. 6 (2024).

Piña, Christy. "Mark Zuckerberg Clarifies Reports About His Bunker in Hawaii: 'That's Just Like a Little Shelter.'" *Hollywood Reporter,* Dec. 21, 2024.

Plummer, Libby. "This Is How Netflix's Top-Secret Recommendation System Works." *Wired,* Aug. 22, 2017.

Ponciano, Jonathan. "Google Billionaire Eric Schmidt Warns of 'National Emergency' if China Overtakes U.S. in AI Tech." *Forbes,* March 7, 2021.

Popper, Karl. *The Poverty of Historicism.* London: Routledge, 2002.

Porter, Theodore. *The Rise of Statistical Thinking, 1820–1900.* Princeton, N.J.: Princeton University Press, 2020.

———. *Trust in Numbers.* Princeton, N.J.: Princeton University Press, 2020.

Protzko, John, et al. "Retraction Note: High Replicability of Newly Discovered Social-Behavioural Findings Is Achievable." *Nature Human Behaviour,* Sept. 24, 2024.

Pugh, Jonathan. "Risky Business: AI and the Future of Insurance." In *AI Morality,* edited by David Edmonds, 30–40. Oxford: Oxford University Press, 2024.

Raworth, Kate. *Doughnut Economics: Seven Ways to Think Like a 21st-Century Economist.* London: Random House Business Books, 2018.

Reich, Rob, Mehran Sahami, and Jeremy M. Weinstein. *System Error: Where Big Tech Went Wrong and How We Can Reboot.* New York: Harper, 2022.

Relman, Eliza. "The 25 Women Who Have Accused Trump of Sexual Misconduct." *Business Insider,* May 1, 2020.

Rescher, Nicholas. *Predicting the Future: An Introduction to the Theory of Forecasting.* Albany: State University of New York Press, 1998.

Ressa, Maria. *How to Stand Up to a Dictator.* London: W. H. Allen, 2023.

Richmond, Shane. "The Printed Book Is Doomed: Here's Why." *Telegraph,* Aug. 4, 2011.

Roberts, Andrew. *Churchill: Walking with Destiny.* London: Penguin, 2019.

Rochberg, Francesca. *The Heavenly Writing: Divination, Horoscopy, and Astronomy in Mesopotamian Culture.* Cambridge, U.K.: Cambridge University Press, 2004.

Rodríguez, Delia. "El hombre más rico del mundo no duerme mejor que tú." *El país,* Feb. 6, 2025.

Romero, Felipe. "Philosophy of Science and the Replicability Crisis." *Philosophy Compass* 14, no. 11 (2019).

Rosa, Hartmut. *The Uncontrollability of the World.* Cambridge, U.K.: Polity Press, 2020.

Rosecrance, Richard N., and Steven E. Miller, eds. *The Next Great War? The Roots of World War I and the Risk of U.S.-China Conflict.* Cambridge, Mass.: MIT Press, 2014.

Rosenthal, Robert, and Kermit L. Fode. "The Effect of Experimenter Bias on the Performance of the Albino Rat." *Behavioral Science* 8, no. 3 (1963): 183–89.

Rothman, Lily. "What Happened to the Car Industry's Most Famous Flop?" *Time,* Nov. 19, 2014.

Runciman, David. *The Handover: How We Gave Control of Our Lives to Corporations, States, and AIs.* London: Profile, 2024.

Rushkoff, Douglas. "The Super-Rich 'Preppers' Planning to Save Themselves from the Apocalypse." *Guardian,* Sept. 4, 2022.

Rutenberg, Jim, Ken Bensinger, and Steve Eder. "The 'Red Wave' Washout: How Skewed Polls Fed a False Election Narrative." *New York Times,* Dec. 31, 2022.

Sackett, Morgan, and Darin Henry. *Seinfeld: How It Began.* 2004.

Sargent, Greg, and Michael Tomasky. " 'Red Wave' Redux: Are GOP Polls Rigging the Averages in Trump's Favor?" *New Republic,* Oct. 23, 2024.

Schaake, Marietje. "Lobbying for Unfettered Innovation Is Bad for Democracy." *Financial Times,* Sept. 24, 2024.

————. *The Tech Coup.* Princeton, N.J.: Princeton University Press, 2024.

Schleifer, Theodore. "Elon Musk Still Holds a Bit of Leverage over President Trump." *New York Times,* June 5, 2025.

Schweizer, Renate. "Unraveling the True Identity of the Brain of Carl Friedrich Gauss." Press release, Max Planck Institute, Oct. 28, 2013.

Scott, A. O. "The Supervillain Is the Hero Now." *New York Times,* Nov. 23, 2024.

Scott, James C. *Seeing Like a State: How Certain Schemes to Improve the Human Condition Have Failed.* New Haven, Conn.: Yale University Press, 1998.

Searcey, Dionne, and Madison Malone Kircher. "Torrent of Hate for Health Insurance Industry Follows CEO's Killing." *New York Times,* Dec. 5, 2024.

Seinfeld, Jerry. *SeinLanguage.* New York: Bantam, 2008.

Seneca. *On the Shortness of Life.* London: Penguin, 1997.

Seth, Anil. *Being You: A New Science of Consciousness.* London: Faber & Faber, 2021.

Shadbolt, Nigel, and Roger Hampson. *As If Human: Ethics and Artificial Intelligence.* New Haven: Yale, 2024.

Shanbhogue, Rachana. "Will the Bubble Burst for AI in 2025, or Will It Start to Deliver?" *Economist,* Nov. 18, 2024.

Shapin, Steven. *The Scientific Revolution.* Chicago: University of Chicago Press, 1998.

Sherden, William A. *The Fortune Sellers: The Big Business of Buying and Selling Predictions.* New York: John Wiley & Sons, 1998.

Shoemaker, David. *Wisecracks: Humor and Morality in Everyday Life.* Chicago: University of Chicago Press, 2024.

Sidgwick, Henry. *The Methods of Ethics.* London: Macmillan, 1907.

Silver, Nate. *On the Edge: The Art of Risking Everything.* London: Allen Lane, 2024.

————. *The Signal and the Noise.* London: Penguin, 2012.

Simmons, Joseph P., Leif D. Nelson, and Uri Simonsohn. "False-Positive Psychology: Undisclosed Flexibility in Data Collection and Analysis Allows Presenting Anything as Significant." *Psychological Science* 22, no. 11 (2011).

Singer, Peter. "The Hinge of History." *Project Syndicate,* Oct. 8, 2021.

Sloan, Dylan, and Tom Maloney. "Elon Musk's Feud with Trump Spurs One of His Worst Wealth Losses Ever." *Bloomberg,* June 5, 2025.

Smith, Ian, and Martin Arnold. "AI Risks Making Some People 'Uninsurable,' Warns UK Financial Watchdog." *Financial Times,* Sept. 19, 2024.

Solal, Isabelle, and Kaisa Snellman. "For Female Founders, Fundraising Only from Female VCs Comes at a Cost." *Harvard Business Review,* Feb. 1, 2023.

Sommerlad, Joe. "Silicon Six Accused of Avoiding Nearly $300Bn in Taxes, New Report Says." *Independent,* April 15, 2025.

Spiegelhalter, David. *The Art of Uncertainty.* London: Pelican, 2024.

Srinivasan, Amia. "Stop the Robot Apocalypse." *London Review of Books,* Sept. 24, 2015.

Stanton, Andrew. "A Year Before CEO Shooting, Lawsuit Alleged UHC Used AI to Deny Coverage." *Newsweek,* Dec. 5, 2024.

Stokel-Walker, Chris. "Revealed: How the UK Tech Secretary Uses ChatGPT for Policy Advice." *New Scientist,* March 13, 2025.

Stone, Brad. "Amazon Erases Orwell Books from Kindle." *New York Times,* July 17, 2009.

Strittmatter, Kai. *We Have Been Harmonised: Life in China's Surveillance State.* Exeter: Old Street Publishing, 2019.

Suetonius. *The Twelve Caesars.* Translated by Robert Graves. London: Penguin Classics, 2007.

Surowiecki, James. *The Wisdom of Crowds.* New York: Vintage, 2005.

Susskind, Daniel. *Growth: A Reckoning.* London: Allen Lane, 2024.

Taleb, Nassim Nicholas. *Antifragile: Things That Gain from Disorder.* New York: Random House, 2012.

———. *The Black Swan: The Impact of the Highly Improbable.* London: Penguin, 2010.

Tau, Byron. *Means of Control: How the Hidden Alliance of Tech and Government Is Creating a New American Surveillance State.* New York: Crown, 2024.

Tetlock, Philip. *Expert Political Judgment.* Princeton, N.J.: Princeton University Press, 2017.

Tett, Gillian. "Can AI Crack Comedy?" *Financial Times,* Aug. 26, 2023.

Thompson, John B. *Merchants of Culture.* London: Polity Press, 2012.

Thomson, Judith Jarvis. "Killing, Letting Die, and the Trolley Problem." *Monist* 59 (1976): 204–17.

Thucydides. *History of the Peloponnesian War.* Translated by Rex Warner. New York: Penguin, 1972.

Thurston, Joshua. "Former Oxford Islamic Scholar Convicted of Rape in Switzerland." *Times,* Sept. 10, 2024.

Tolstoy, Leo. *The Gospel in Brief.* Translated by Dustin Condren. London: Harper Perennial, 2011.

Torres, Émile P. "Fraud, Lies, Exploitation, and Eugenic Fantasies." *TruthDig,* Nov. 9, 2023.

The Trial: The DOJ's Suit to Block Penguin Random House's Acquisition of Simon & Schuster. Bronxville, N.Y.: Publishers Lunch, 2022.

Trotta, Roberto. *Starborn.* London: Basic Books, 2023.

Tsang, Eric W. K. "Toward a Scientific Inquiry into Superstitious Business Decision-Making." *Organization Studies* 25, no. 6 (2004).

Ugwu, Reggie. "It's Their Content, You're Just Licensing It." *New York Times,* April 4, 2023.

Vallor, Shannon. *The AI Mirror: How to Reclaim Our Humanity in the Age of Machine Thinking.* Oxford: Oxford University Press, 2024.

van der Horst, Peter W. "Ancient Jewish Bibliomancy." In *Divination: Oracles & Omens,* edited by Michelle Aroney and David Zeitlyn. Oxford: Bodleian Library Publishing, 2024.

Véliz, Carissa. "Chatbots Shouldn't Use Emojis." *Nature* 615 (2023): 375.

———. *The Ethics of Privacy and Surveillance.* London: Oxford University Press, 2024.

———. "If AI Is Predicting Your Future, Are You Still Free?" *Wired,* Dec. 27, 2021.

———. "If You Want to Do the Most Good, Maybe You Shouldn't Work for Wall Street." *Practical Ethics in the News,* July 9, 2015.

———. "Moral Zombies: Why Algorithms Are Not Moral Agents." *AI & Society* 36 (2021): 487–97.

———. *Privacy Is Power.* London: Bantam Press, 2020.

———. "Three Things Digital Ethics Can Learn from Medical Ethics." *Nature Electronics* 2 (2019): 316–18.

———. "Why LaMDA Is Nothing Like a Person. Language Models Are No More Sentient than Your Reflection in the Mirror." *Slate,* June 21, 2022.

Vinay, Rasita, Holger Baumann, and Nikola Biller-Andorno. "Ethics of ICU Triage During Covid-19." *British Medical Bulletin* 138, no. 1 (2021): 5–15.

Vincent, James. *Beyond Measure: The Hidden History of Measurement.* London: Faber & Faber, 2022.

———. "Instagram Internal Research: 'We Make Body Image Issues Worse for One in Three Teen Girls.'" *Verge,* Sept. 15, 2021.

Visram, Talib. "The Truth About Elon Musk, Sam Bankman-Fried, and Effective Altruism." *Fast Company,* Sept. 14, 2022.

Volpi, Jorge. *La invención de todas las cosas: Una historia de la ficción.* Madrid: Alfaguara, 2024.

Vowels, Laura M. "Are Chatbots the New Relationship Experts? Insights from Three Studies." *Computers in Human Behavior: Artificial Humans* 2, no. 2 (2024).

Ward, Jacob. *The Loop.* New York: Hachette, 2022.

Watson, Clare. "Many Researchers Say They'll Share Data—but Don't." *Nature,* June 21, 2022.

Watson, Gary. "Moral Agency." In *The International Encyclopedia of Ethics,* edited by Hugh LaFollette. Malden, Mass.: Wiley-Blackwell, 2013.

Weber, Max. *Economy and Society.* Berkeley: University of California Press, 1978.

Wee, Sui-Lee. "China Uses DNA to Track Its People, with the Help of American Expertise." *New York Times,* Feb. 21, 2019.

Weintraub, Karen. "Chatbots Are Accurate, Show More Empathy than Doctors in Answering Questions, Study Finds." *USA Today,* May 1, 2023.

Weiser, Benjamin, and Nate Schweber. "The ChatGPT Lawyer Explains Himself." *New York Times,* June 8, 2023.

Wenar, Leif. "The Deaths of Effective Altruism." *Wired,* March 27, 2024.

Whitney, Craig R. "In France, a Citizen Turns 120." *New York Times,* Feb. 22, 1995.

Wiggers, Kyle. "Anthropic Publishes the 'System Prompts' That Make Claude Tick." *TechCrunch,* Aug. 26, 2024.

Williams, Bernard. *Making Sense of Humanity and Other Philosophical Essays.* Cambridge, U.K.: Cambridge University Press, 1995.

———. *Morality.* Cambridge, U.K.: Cambridge University Press, 2019.

Williams, Joan C., and Kate Massinger. "How Women Are Harassed out of Science." *Atlantic,* July 25, 2016.

Williams, Thomas Chatterton. "The People Who Don't Read Books." *Atlantic,* Jan. 25, 2023.

Wilson, Jason. "Private Firms Provide Software and Information to Police, Documents Show." *Guardian,* Oct. 15, 2020.

Wu, Tim. *The Age of Extraction: How Tech Platforms Conquered the Economy and Threaten Our Future Prosperity.* New York: Alfred A. Knopf, 2025.

Wylie, Christopher. *Mindf*ck: Inside Cambridge Analytica's Plot to Break the World.* London: Profile, 2019.

Yates, Kit. *How to Expect the Unexpected: The Science of Making Predictions and the Art of Knowing When Not To.* London: Quercus, 2023.

INDEX

ABOUT THE AUTHOR

Carissa Véliz is an associate professor at the Faculty of Philosophy and the Institute for Ethics in AI, as well as a tutorial fellow at Hertford College, at the University of Oxford. Her first book, *Privacy Is Power,* was an *Economist* Book of the Year and has been published in seven languages. She is also the author of *The Ethics of Privacy and Surveillance* and the editor of *The Oxford Handbook of Digital Ethics.* She advises companies and governments around the world on ethics, privacy, and AI.